KU-338-697

WITHDRAWN

MOBILE TELECOMMUNICATIONS PROTOCOLS FOR DATA NETWORKS

MOBILE TELECOMMUNICATIONS PROTOCOLS FOR DATA NETWORKS

Anna Hać
University of Hawaii at Manoa, Honolulu

JOHN WILEY & SONS, LTD

Telephone (+44) 1243 779777

Email (for orders and customer service enquiries): cs-books@wiley.co.uk
Visit our Home Page on www.wileyeurope.com or www.wiley.com

Other Wiley Editorial Offices

John Wiley & Sons Inc., 111 River Street, Hoboken, NJ 07030, USA

Jossey-Bass, 989 Market Street, San Francisco, CA 94103-1741, USA

Wiley-VCH Verlag GmbH, Boschstr. 12, D-69469 Weinheim, Germany

John Wiley & Sons Australia Ltd, 33 Park Road, Milton, Queensland 4064, Australia

John Wiley & Sons (Asia) Pte Ltd, 2 Clementi Loop #02-01, Jin Xing Distripark, Singapore 129809

John Wiley & Sons Canada Ltd, 22 Worcester Road, Etobicoke, Ontario, Canada M9W 1L1

British Library Cataloguing in Publication Data

A catalogue record for this book is available from the British Library

ISBN 0-470-85056-6

Typeset in 10/12pt Times by Laserwords Private Limited, Chennai, India
Printed and bound in Great Britain by TJ International, Padstow, Cornwall
This book is printed on acid-free paper responsibly manufactured from sustainable forestry
in which at least two trees are planted for each one used for paper production.

Contents

Preface

Mobile telecommunications emerged as a technological marvel allowing for access to personal and other services, devices, computation and communication, in any place and at any time through effortless plug and play. This brilliant idea became possible as the result of new technologies developed in the areas of computers and communications that were made available and accessible to the user.

This book describes the recent advances in mobile telecommunications and their protocols. Wireless technologies that expanded to a wide spectrum and short-range access allow a large number of customers to use the frequency spectrum when they need it. Devices are used to communicate with the expanded network. Software systems evolved to include mobile agents that carry service information that is compact enough to be implemented in the end user devices.

The area of mobile telecommunications has been growing rapidly as new technologies emerge. Mobile users are demanding fast and efficient connections that support data applications. Extending wireless access to the applications requires creating mobile agents, systems, and platforms to implement service configuration. Wireless Local Area Networks (LANs) supporting a growing number of users and applications require wideband wireless local access, wireless protocols, and virtual LANs. Wireless applications require protocols and architecture supporting these applications. Wireless connection has to be provided by the networks and protocols. Mobile networks must function efficiently by using their protocols, performing routing and handoff for mobile users.

This book focuses on the newest technology for mobile telecommunications supporting data applications. The book provides a real application-oriented approach to solving mobile communications and networking problems. The book addresses a broad range of topics from mobile agents and wireless LANs to wireless application protocols, wireless architecture, and mobile networks.

This book proposes a comprehensive design for mobile telecommunications including mobile agents, access networks, application protocols, architecture, routing, and handoff. For mobile users and data applications, these are new networking and communications solutions, particularly for the LAN environment. The book describes the aspects of mobile telecommunications for applications, networking, and transmission. Additionally, it introduces and analyzes architecture and design issues in mobile communications and networks.

The book is organized into 12 chapters. The first seven chapters describe applications, their protocols and mobile and wireless network support for them. Chapters 8 through 12 describe architecture of mobile and wireless networks, their protocols, and quality-of-service (QoS) issues.

The goal of this book is to explain how to support modern mobile telecommunications, which evolve toward value-added, on-demand services, in which the need for communication becomes frequent and ongoing, and the nature of the communication becomes more complex. Mobile agents are used to enable on-demand provision of customized services. Examples of mobile agent-based service implementation, middleware, and configuration are introduced.

Mobile applications are supported by wireless LANs. Virtual LANs provide support for workgroups that share the same servers and other resources over the network.

Orthogonal Frequency Division Multiplex (OFDM) allows individual channels to maintain their orthogonality, or distance, to adjacent channels. This technique allows data symbols to be reliably extracted and multiple subchannels to overlap in the frequency domain for increased spectral efficiency. The IEEE 802.11 standards group chose OFDM modulation for wireless LANs operating at bit rates up to 54 Mb s^{-1} at 5 GHz.

Wideband Code Division Multiple Access (WCDMA) uses 5-MHz channels and supports circuit and packet data access at 384 kb s^{-1} nominal data rates for macrocellular wireless access. WCDMA provides simultaneous voice and data services. WCDMA is the radio interface technology for Universal Mobile Telecommunications System (UMTS) networks.

Mobile applications and wireless LANs use wireless protocols. A Media Access Control (MAC) protocol for a wireless LAN provides two types of data-transfer Service Access Points (SAP): network and native. The network SAP offers an access to legacy network protocols [e.g., IP (Internet Protocol)]. The native SAP provides an extended service interface that may be used by custom network protocols or user applications capable of fully exploiting the protocol-specific QoS parameters within the service area.

Limitations of power, available spectrum, and mobility cause wireless data networks to have less bandwidth and more latency than traditional networks, as well as less connection stability than other network technologies, and less predictable availability.

Mobile devices have a unique set of features that must be exposed into the World Wide Web (WWW) in order to enable the creation of advanced telephony services such as location-based services, intelligent network functionality, including integration into the voice network, and voice/data integration.

The Wireless Application Protocol (WAP) architecture provides a scalable and extensible environment for application development for mobile communication devices. The WAP protocol stack has a layered design, and each layer is accessible by the layers above and by other services and applications. The WAP layered architecture enables other services and applications to use the features of the WAP stack through a set of well-defined interfaces. External applications can access the session, transaction, security, and transport layers directly.

The network architecture supporting wireless applications includes Wireless Applications Environment (WAE), Wireless Telephony Application (WTA), and WAP Push framework. The WAE architecture is designed to support mobile terminals and network applications using different languages and character sets.

WTA is an application framework for telephony services. The WTA user agent has the capabilities for interfacing with mobile network services available to a mobile telephony device, that is, setting up and receiving phone calls.

The WAP Push framework introduces a means within the WAP effort to transmit information to a device without a previous user action. In the client/server model, a client requests a service or information from a server, which transmits information to the client. In this pull technology, the client pulls information from the server.

Extensible Markup Language (XML) is an application profile or restricted form of the Standard Generalized Markup Language (SGML). XML describes a class of data objects called *XML documents* and partially describes the behavior of computer programs that process them. Resource Description Framework (RDF) can be used to create a general, yet extensible, framework for describing user preferences and device capabilities. This information can be provided by the user to servers and content providers. The servers can use this information describing the user's preferences to customize the service or content provided.

A Composite Capability/Preference Profile (CC/PP) is a collection of the capabilities and preferences associated with the user and the agents used by the user to access the World Wide Web (WWW). These user agents include the hardware platform, system software, and applications used by the user.

In a wireless LAN, the connection between the client and the user exists through the use of a wireless medium such as Radio Frequency (RF) or Infrared (IR) communications. The wireless connection is most usually accomplished by the user having a handheld terminal or a laptop computer that has an RF interface card installed inside the terminal or through the PC (personal computer) card slot of the laptop. The client connection from the wired LAN to the user is made through an Access Point (AP) that can support multiple users simultaneously. The AP can reside at any node on the wired network and performs as a gateway for wireless users' data to be routed onto the wired network.

A wireless LAN is capable of operating at speeds in the range of 1 or 2, or 11 Mbps depending on the actual system. These speeds are supported by the standard for wireless LAN networks defined by the international body, the IEEE.

The network communications use a part of the radio spectrum that is designated as license-free. In this band, of 2.4 to 2.5 GHz, the users can operate without a license when they use equipment that has been approved for use in this license-free band. The 2.4-GHz band has been designated as license-free by the International Telecommunications Union (ITU) and is available for use, license-free in most countries in the world.

The ability to build a dynamically scalable network is critical to the viability of a wireless LAN as it will inevitably be used in this mode. The interference rejection of each node will be the limiting factor to the expandability of the network and its user density in a given environment.

In *ad hoc* networks, all nodes are mobile and can be connected dynamically in an arbitrary manner. All nodes of these networks behave as routers and take part in discovery and maintenance of routes to other nodes in the network.

An *ad hoc* network is a collection of mobile nodes forming a temporary network without the aid of any centralized administration or standard support services available in conventional networks.

Ad hoc networks must deal with frequent changes in topology. Mobile nodes change their network location and link status on a regular basis. New nodes may unexpectedly join the network or existing nodes may leave or be turned off. *Ad hoc* routing protocols must minimize the time required to converge after the topology changes.

The *ad hoc* routing protocols can be divided into two classes: table-driven and on-demand routing on the basis of when and how the routes are discovered. In table-driven routing protocols, consistent and up-to-date routing information to all nodes is maintained at each node, whereas in on-demand routing, the routes are created only when desired by the source host.

When the mobile end user moves from one AP to another AP, a handoff is required. When the handoff occurs, the current QoS may not be supported by the new data path. In this case, a negotiation is required to set up new QoS. Since a mobile user may be in the access range of several APs, it will select the AP that provides the best QoS. During the handoff, an old path is released and then a new path is established. Connection rerouting schemes must exhibit low handoff latency, maintain efficient routes, and limit disruption to continuous media traffic while minimizing reroute updates to the network switches and nodes.

Basically, there are three connection rerouting approaches: full connection establishment, partial connection re-establishment, and multicast connection re-establishment.

In the wireless Asynchronous Transfer Mode (ATM) network, a radio access layer provides high-bandwidth wireless transmission with appropriate medium access control and data link control. A mobile ATM network provides base stations (access points) with appropriate support of mobility-related functions, such as handoff and location management.

QoS-based rerouting algorithm is designed for the two-phase interswitch handoff scheme for wireless ATM networks. Path extension is used for each inter-switch handoff, and path optimization is invoked when the handoff path exceeds the delay constraint or maximum path extension hops constraint. The path optimization schemes include combined QoS-based, delay-based, and hop-based path rerouting schemes.

The content of the book is organized into 12 chapters as follows:

Chapter 1 introduces mobile agents and presents platforms and systems to implement agent-based services in the network. Chapter 2 describes mobile agent-based service implementation. Mobile agent-based middleware and service configuration are introduced. Mobile agent implementation is discussed.

Chapter 3 describes wireless LANs, introduces virtual LANs, and presents wideband wireless local access. Chapter 4 describes wireless protocols.

Protocols for wireless applications are studied in Chapter 5. Wireless applications and devices are discussed and wireless application protocol is introduced. Network architecture supporting wireless applications is presented in Chapter 6. Extensible markup language, resource description framework, and composite capability/preference profile are described in Chapter 7.

Architecture of wireless LANs is studied in Chapter 8. The protocols supporting mobile communications, IEEE 802.11 and Bluetooth, are described.

Routing protocols in mobile and wireless networks are presented in Chapter 9. Handoff in mobile networks is described in Chapter 10. Signaling traffic in wireless networks is studied in Chapter 11. Chapter 12 presents a two-phase combined handoff scheme in wireless networks.

About the Author

Anna Hać received her M.S. and Ph.D. degrees in Computer Science from the Department of Electronics, Warsaw University of Technology, Poland, in 1977 and 1982, respectively.

She is a professor in the Department of Electrical Engineering, University of Hawaii at Manoa, Honolulu. During her long and successful academic career, she has been a visiting scientist at Imperial College, University of London, England, a postdoctoral fellow at the University of California at Berkeley, an assistant professor of electrical engineering and computer science at The Johns Hopkins University, a member of the technical staff at AT&T Bell Laboratories, and a senior summer faculty fellow at the Naval Research Laboratory.

Her research contributions include system and workload modeling, performance analysis, reliability, modeling process synchronization mechanisms for distributed systems, distributed file systems, distributed algorithms, congestion control in high-speed networks, reliable software architecture for switching systems, multimedia systems, and wireless networks.

She has published more than 130 papers in archival journals and international conference proceedings and is the author of a textbook *Multimedia Applications Support for Wireless ATM Networks* (2000).

She is a member of the Editorial Board of the *IEEE Transactions on Multimedia* and is on the Editorial Advisory Board of Wiley's *International Journal of Network Management*.

1

Mobile agent platforms and systems

Advanced service provisioning allows for rapid, cost-effective service deployment. Modern mobile telecommunications evolve towards value-added, on-demand services in which the need for communication becomes frequent and ongoing, and the nature of the communication becomes more complex. The services of the future will be available 'a la carte', allowing subscribers to receive content and applications when they want it.

Introducing Mobile Agents (MAs) within the network devices, Mobile Stations (MSs), and Mobile Switching Centers (MSCs) provides the necessary flexibility into the network and enhanced service delivery. MAs enable on-demand provision of customized services via dynamic agent downloading from the provider system to the customer system or directly to the network resources. MAs have the capability to migrate between networks, to customize for the network, and to decentralize service control and management software by bringing control and managements agents as close as possible to the resources.

MAs can be used in mobile networks to support advanced service provisioning, as well as for personal communication, for mobility, and to support Virtual Home Environment (VHE). The VHE agent enables individually subscribed and customized services to follow their associated users to wherever they roam.

1.1 MOBILE AGENT PLATFORMS

Mobile Agent Technology (MAT) uses interworking between Mobile Agent Platforms (MAPs). Several MAPs are based on Java. These platforms are Grasshopper, Aglets, Concordia, Voyager, and Odyssey.

Each MAP has a class library that allows the user to develop agents and applications. The core abstractions are common to most platforms since they are inherent in the MA paradigm. These abstractions include agents, hosts, entry points, and proxies.

- *Agents*: In each platform, a base class provides the fundamental agent capability. In some platforms this base class is used for all agents (static and mobile) while in others there are two separate classes.
- *Hosts*: The terms *hosts*, *environments*, *agencies*, *contexts*, *servers*, and *AgentPlaces* are used to refer to the components of the framework that must be installed at a computer node and that provide the necessary runtime environment for the agents to execute.
- *Entry points*: The agents have to save the necessary state information to member variables, allowing the entry point method to proceed depending on the state of the computation. Platforms may have one or multiple entry points.
- *Proxies*: The proxy is a representative that an MA leaves when migrating from a node, and it can be used to forward messages or method invocations to an MA in a location-independent manner. Platforms may implement proxies in different ways. A significant difference is whether the arbitrary methods of an agent can be called remotely through the proxy. Platforms that support this functionality provide a utility that parses a MA's class and creates a corresponding proxy. In platforms where arbitrary Remote Method Invocation (RMI) through a proxy is not supported, the proxy object provides only a uniform, generic method to send messages, and therefore no proxy-generation utility is required.

1.1.1 Grasshopper

The Grasshopper platform consists of a number of agencies (hosts) and a Region Registry (a network-wide database of host and agent information) remotely connected *via* an Object Request Broker (ORB). Agencies represent the runtime environments for MAs. Several agencies can be grouped into one region represented by a region registry.

Remote interactions between the components of the Distributed Agent Environment (DAE) are performed *via* an ORB. The Grasshopper's Communication Service is a part of each agency and region registry. The Grasshopper supports the following protocols: plain sockets (with or without Secure Socket Layer, SSL), Common Object Request Broker Architecture (CORBA) Internet Inter – ORB Protocol (IIOP), and RMI – with or without SSL. Support for more protocols can be integrated into the communication service.

The Grasshopper platform conforms to the Object Management Group's (OMG) Mobile Agent System Interoperability Facility (MASIF) standard.

1.1.2 Aglets

Aglets (Agent applets) were developed by the IBM Tokyo Research Laboratory. The Aglets class library provides an Application Programming Interface (API) that facilitates the encoding of complex agent behavior. Particularly, the way the behavior of the base Aglet class is extended resembles the way Web applets are programmed. Aglets can cooperate with web browsers and Java applets.

The communication API used by Aglets is derived from MASIF standard. The default implementation of the API is the Agent Transfer Protocol (ATP). ATP is an application level protocol based on TCP and modeled on the Hypertext Transfer Protocol (HTTP) for transmitting messages and MAs between the networked computers in which the hosts reside. The core Aglet runtime is independent of the transport protocol and accesses ATP through a well-defined interface. Aglets use an interface, derived from MASIF standard, for the internal communication between the runtime core and the communication system, but do not export this interface as an external CORBA interface. The latest version of Aglets supports ATP and RMI. A CORBA IIOP–based transport layer will be provided in the future release of Aglets.

1.1.3 Concordia

Concordia was developed by Mitsubishi Electric Information Technology Center, USA. The main component of the Concordia system is the Concordia server that provides for the necessary runtime support. The server consists of components integrated to create MA framework.

Concordia uses TCP/IP communication services. The communication among agents and their migration employs Java's RMI, where standard sockets are replaced by secure sockets (SSL).

1.1.4 Voyager

Voyager developed by ObjectSpace is a Java-based MA system. Voyager relies exclusively on the services of its supporting ORB. The core functionality of an ORB is to facilitate interobject communication by shuttling messages to and from remote objects and instantiating persistent distributed objects. Voyager's ORB can facilitate only Java objects, and this is not an OMG-compatible ORB.

Features supported by the Voyager's ORB include migration of both agents and arbitrary Java object (a feature that does not exist in other MAPs), the ability to remote-enable (instantiate) a class, remote execution of static methods, multicast messaging, synchronous messages, and time-dependent garbage collection. ObjectSpace has implemented hooks in the Voyager to support interworking with other ORBs.

1.1.5 Odyssey

Odyssey is a Java-based MAP implemented by General Magic. Odyssey uses Java's RMI for communication between Agents. The transport mechanism used for Agent migration can be CORBA IIOP, Distributed Component Object Model (DCOM), or RMI. Agents cannot call remotely the methods of other Agents but can engage with them in a meeting.

1.2 MULTIAGENT SYSTEMS

Agent-based technology offers a solution to the problem of designing efficient and flexible network management strategies. The OMG has produced the MASIF, which focuses on mobile agent (object) technology, in particular, allowing for the transfer of agents code and state between heterogeneous agent platforms.

The Intelligent Network (IN) was developed to introduce, control, and manage services rapidly, cost effectively, and in a manner not dependent on equipment and software from particular equipment manufactures. The architecture of an IN consists of the following node types: Service Switching Points (SSPs), Service Control Points (SCPs), Service Data Points (SDPs), and Intelligent Peripherals (IP). These nodes communicate with each other by using a Signaling System No. 7 (SS7) network. SSPs facilitate end user access to services by using trigger points for detection of service access codes. SCPs form the core of the architecture; they receive service requests from SSPs and execute the service logic. SCPs are assisted by SDPs, which store service/customer related data, and by IPs, which provide services for interaction with end users (e.g., automated announcements or data collection).

IN overloads occur when the load offered to one or more network resources (e.g., SCP processors) exceeds the resource's maximum capacity. Because of the central role played by the SCP, the overall goal of most IN load control mechanisms is to protect SCP processors from overload. The goal is to provide customers with high service availability and acceptable network response times, even during periods of high network loading. Load control mechanisms are designed to be

- *efficient* – keeping SCP utilization high at all times;
- *scalable* – suited to all networks, regardless of their size and topology;
- *responsive* – reacting quickly to changes in the network or offered traffic levels;
- *fair* – distributing system capacity among network users and service providers in a manner deemed fair by the network operator;
- *stable* – avoiding fluctuations or oscillations in resources utilization;
- *simple* – in terms of ease of implementation.

The majority of IN load control mechanisms are node-based, focusing on protecting individual nodes in the network (typically SCPs) from overload. Jennings *et al.* argue that node-based mechanisms cannot alone guarantee that desired Quality of Service (QoS) levels are consistently achieved. The following observations support this viewpoint:

- Most currently deployed node-based mechanisms were designed for standard telephony traffic patterns. Present and future INs support a large number of heterogeneous services, each exhibiting changing traffic characteristics that cannot be effectively controlled by using node-based techniques.
- Existing node-based overload protection mechanisms serve to protect individual nodes only and may cause the propagation of traffic congestion, resulting in adverse effects on the service completion rates of the network as a whole.
- Typically node-based mechanisms do not interact effectively with the protection mechanisms that are incorporated into the signaling networks that carry information between the nodes in a network.
- Node-based controls typically focus on SCP protection only.
- Telecommunications equipment manufactures implement node-based mechanisms on a proprietary basis. This can lead to difficulties in effectively controlling traffic in INs that contain heterogeneous types of equipment.

While flexible and adaptable network-based load control mechanisms can be implemented by using standard software engineering techniques, Jennings *et al.* argue that there are many advantages of adopting an agent-based approach:

- *Methodology*: The agent paradigm encourages an information-centered approach to application development; thus it provides a useful methodology for the development of control mechanisms that require manipulation of large amounts of data collected throughout the network.
- *Agent communication languages*: Advanced communication languages allow agents to negotiate in advance the semantics of future communications. This is not present in traditional communications protocols and can be used in mechanisms that adapt to dynamic network environments in which, for instance, traffic patterns change as a result of the introduction or withdrawal of services.
- *Adaptivity*: The agents adaptive behavior allows them to learn about the normal state of the network and better-judge their choice of future actions.
- *Openness*: Agents can exchange data and apply it in different ways to achieve a common goal. This means that equipment manufacturers can develop load control agents for their own equipment, but these agents can still communicate with agents residing in other equipment types.
- *Scalability*: The agent approach allows for increased scalability to larger networks. For instance, an agent associated with a recently introduced piece of equipment can easily incorporate itself into the agent community and learn from the other agents the range of parameters that it should use for its load control algorithm.
- *Robustness*: Agents typically communicate asynchronously with each other and thus are not dependent on the prompt delivery of interagent messages. The ability to act even during interrupted communications (e.g., due to overload or network failures) is a desirable attribute of a load control mechanism.

1.2.1 Agent-based load control strategies

The goal of the agent-based load control strategies is to allocate resources to the arriving user service requests in an optimal way. There are three classes of agents that carry out the tasks necessary to allocate IN resources in this optimal way:

- *QUANTIFIER* agents that monitor and predict the load and performance of SCP processors (and possibly other IN resources) and report this information to the other agents;
- *DISTRIBUTOR* agents that maintain an overview of the load and resource status in the entire network and can play a controlling and supervisory role in resource allocation;
- *ALLOCATOR* agents that are associated with SSPs. They form a view of the load situation in the network and the possibility of resource overload, based on their own predictive algorithms and information received from the other agents. If these agents perceive a danger of overload of resources, they throttle service requests on a priority basis.

The allocation of the processing capacity of a number of SCPs between requests for a number of IN service types can be controlled by strategies using the agents: QUANTIFIERS, DISTRIBUTORS, and ALLOCATORS. A simple network containing SSPs and

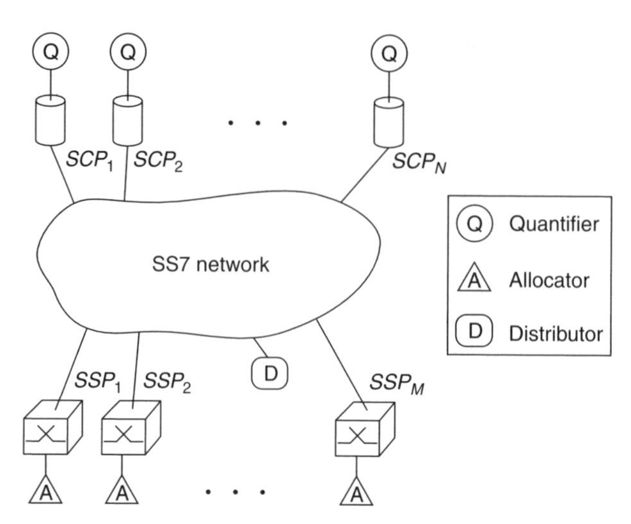

Figure 1.1 Agent-based load control strategy.

SCPs, each supporting all service types, is shown in Figure 1.1 and is used to describe agent-based load control strategies.

Computational markets, as applied to resource allocation problems, are generally implementations of the General Equilibrium Theory, developed in the field of microeconomics, whereby agents in the market set prices and create bids for resources, on the basis of demand-and-supply functions. Once equilibrium has been computed from the bids of all the agents, the resources are allocated in accordance with the bids and the equilibrium prices. The search for the market equilibrium can be implemented so that the customer and producer submit bids to an auctioneer. From these bids, the auctioneer updates its information and requests new bids in an iterative fashion. Once the market equilibrium has been found, the allocation of goods is performed in accordance with the bids and market prices.

In the market strategy, load control is carried out by means of tokens, which are sold by MB-QUANTIFIER agents (MB indicates that the agent implements part of a market-based strategy) of providers (SCP) and bought by MB-ALLOCATOR agents of customers (SSP). The amount of tokens sold by an SCP controls the load offered to it, and the amount of tokens bought by an SSP determines how many IN service requests it can accept. Trading of tokens in an auction is carried out so that the common benefit is maximized.

All SSPs contain a number of pools and tokens, one for each SCP and service class pairing. Each time an SSP feeds an SCP with a service request, one token is removed from the relevant pool. An empty pool indicates that the associated SCP cannot accept more requests of that type from the SSP. Tokens are periodically assigned to pools by an MB-DISTRIBUTOR, which uses an auction algorithm to calculate token allocations. Auctions are centrally implemented by an MB-DISTRIBUTOR using bids received in the form of messages every interval from all the MB-ALLOCATORS and MB-QUANTIFIERS in the network.

MB-QUANTIFIER bids consist of the unclaimed processing capability for the coming interval and the processing requirements for each service class. MB-ALLOCATOR bids consist of the number of expected IN service requests over the next interval for each service class. These values are set to the numbers that arrived in the previous interval as they are assumed to be reasonably accurate estimates.

The objective of the auction process is to maximize expected network profit over the next interval by maximizing the increase in expected marginal utility, measured as marginal gain over cost, for every token issued. The expected marginal gain associated with allocating an additional token to an MB-ALLOCATOR is defined as the profit associated with consuming it times the probability that it will be consumed over the auction interval. The expected marginal cost associated with issuing a token from an MB-QUANTIFIER is defined as the ratio between the processing time consumed and the remaining processing time. On the basis of these values, the MB-DISTRIBUTOR implements a maximization algorithm that is iterated to allocate all the available tokens. Tokens are typically allocated to MB-ALLOCATORS with higher bids (i.e., those that expect greater number of requests for service sessions that result in high profits) in preference to those with lower bids.

The operation of the auction algorithm in which there is only one service class supported by the network is shown in Figure 1.2. In the first step, that is, Bid Submission, MB-QUANTIFIERS and MB-ALLOCATORS submit their bids to the MB-DISTRIBUTOR, which then executes the second step, that is, Auction Process. In this figure, dark circles represent tokens, whereas light circles represent token requests; the auction algorithm assigns tokens to token requests. Once the auction is completed, in the third step the

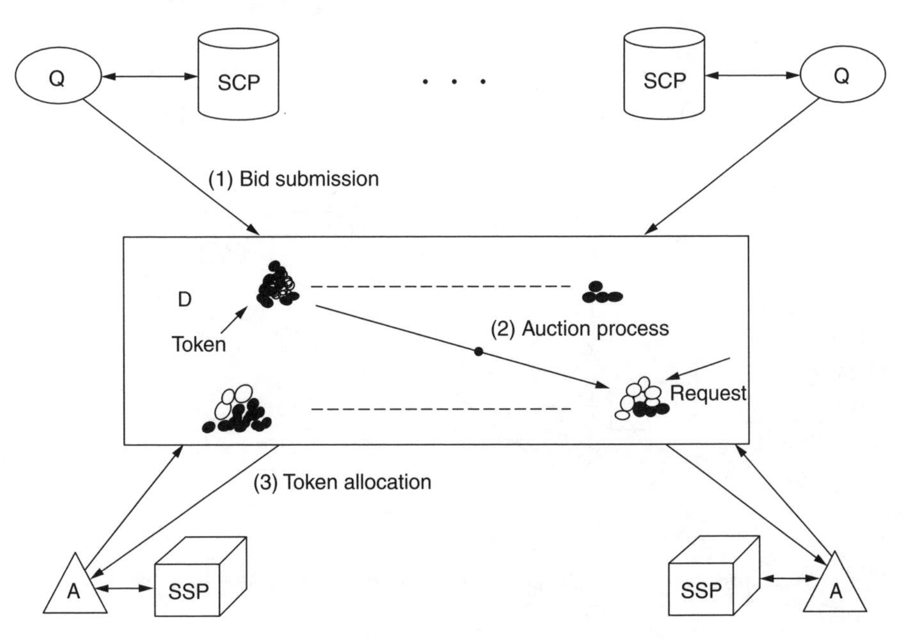

Figure 1.2 Auction algorithm with one service class in cooperative market strategy.

values of token assignments are reported to the MB-ALLOCATORS, which use them to admit service requests in the next time period.

The result of the auction process is that tokens are allocated to balance the arriving traffic load across all SCPs, subject to maximizing the overall network profit.

The following load control strategy is based on Ant Colony Optimization, which is the application of approaches based on the behavior of real ant colonies to optimization problems. The operation of ant-based IN load control strategy is shown in Figure 1.3.

At intervals of length T, a mobile agent AB-ANT, where AB indicates ant-based strategy, is generated for every service type at every SSP in the network and sent to a selected SCP. Each SSP maintains pheromone tables for each service type, which contain entries for all the SCPs in the network. These entries are the normalized probabilities, P_i for choosing SCP_i as the destination for an AB-ANT. The destination SCP of an AB-ANT is selected using the information in the pheromone table following either the normal scheme or the exploration scheme. The scheme used is selected at random, but with the probability of using the normal scheme much higher than the exploration scheme.

In the normal scheme, the SCP is selected randomly, the probability of picking SCP_i being the probability P_i indicated in the pheromone table. In the exploration scheme, the SCP is also selected randomly, and the probabilities of selecting all the SCPs are equal. The purpose of the exploration scheme is to introduce an element of noise into the system so that more performant SCPs can be found.

AB-ANTS travel to the designated SCP, where they interact with the local AB-QUANTIFIER agent and then return to their originating SSP. They also keep track of the time they have spent traversing the network. AB-ANTS arriving at the SCP request information from the AB-QUANTIFIER on the currently expected average processing

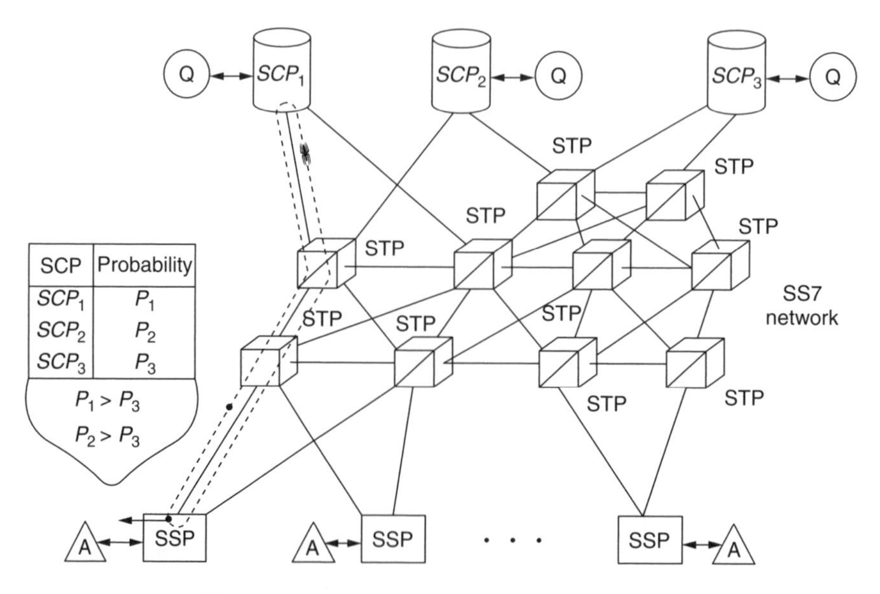

Figure 1.3 Ant-based IN load control strategy.

times for the service type of interest. Processing times reported are the processing time for the initial message of the service session and the sum of the processing times for all other messages. The separation between the processing times for the initial and subsequent messages is used to highlight the importance of the time spent processing the initial message, by which time the service user would not have received any response from the network. Reported processing times include those incurred in accessing information from databases, which may be held in SDPs in other parts of the network.

Upon return to the SSP, the AB-ANT passes its gathered information to the AB-ALLOCATOR, which then updates the pheromone table entries for its service type, using the following formula.

$$P_i = \frac{P_i + \Delta p}{1 + \Delta p}$$

where i indicates the visited SCP, and P_i is the probability of choosing SCP_i. The probability P_j of choosing SCP_j is

$$P_j = \frac{P_j}{1 + \Delta p}, \quad j \in [1, N], \; j \neq i$$

with

$$\Delta p = \frac{a}{t_1} + \frac{b}{t_2} + \frac{c}{t_3} + \frac{d}{t_4} + e$$

where a, b, c, d, and e are constants; t_1 is time-elapsed traveling SSP $\rightarrow$ SCP; t_2 is expected mean SCP processing time for initial message; t_3 is expected mean SCP processing time for subsequent messages; t_4 is time-elapsed traveling SCP $\rightarrow$ SSP.

The values of a, b, c, and d represent the relative importance the AB-ALLOCATOR gives to each of the four measurements. Requests for service are routed to the SCP that has the current highest priority value in the service's pheromone table. Figure 1.3 illustrates that in normal load conditions the operation of the strategy will mean that SCPs with closer proximity to a source are more likely to be chosen as the destination for service requests, the reason being that the delays AB-ANTS experience in traveling to and from them are lower than for other SCPs.

1.3 SUMMARY

Each MAP has a class library that allows the user to develop agents and applications. The core abstractions are common to most platforms since they are inherent in the MA paradigm. These abstractions include agents, hosts, entry points, and proxies.

Agent-based technology offers a solution to the problem of designing efficient and flexible network management strategies. The OMG has produced the MASIF standard, which focuses on MA (object) technology, in particular, allowing for the transfer of agents code and state between heterogeneous agent platforms.

Load control mechanisms are designed to be efficient, scalable, responsive, fair, stable, and simple.

The majority of IN load control mechanisms are node-based, focusing on protecting individual nodes in the network (typically SCPs) from overload. Node-based mechanisms cannot alone guarantee that desired QoS levels are consistently achieved.

Flexible and adaptable network-based load control mechanisms can be implemented by using standard software engineering techniques. There are many advantages of adopting an agent-based approach, which include methodology, agent communication languages, adaptivity, openness, scalability, and robustness.

PROBLEMS TO CHAPTER 1

Mobile agent platforms and systems

Learning objectives

After completing this chapter, you are able to

- demonstrate an understanding of MAs;
- discuss what is meant by MA platforms;
- explain what agent-based technology is;
- demonstrate an improvement to network load control mechanisms;
- explain what an intelligent network is;
- discuss the node-based and agent-based approach.

Practice problems

1.1: What are the core abstractions common to most platforms in the mobile agent paradigm?
1.2: What are the requirements for the load control mechanisms?
1.3: What are the advantages of using an agent-based approach to load control?

Practice problem solutions

1.1: The core abstractions common to most platforms in the mobile agent paradigm include agents, hosts, entry points, and proxies.
1.2: Load control mechanisms are designed to be efficient, scalable, responsive, fair, stable, and simple.
1.3: The advantages of adopting an agent-based approach to load control include methodology, agent communication languages, adaptivity, openness, scalability, and robustness.

2

Mobile agent-based service implementation, middleware, and configuration

There are two agents groups: Intelligent Agents and Mobile Agents (MAs). Intelligent Agents have the ability to learn and react. MAs can migrate between different hosts, execute certain tasks, and collaborate with other agents.

In the Intelligent Network (IN) architecture, the control of the network resources is performed by the signaling plane, whereas the service creation, deployment, and provisioning is performed by the service plane. This separation allows introduction of new services and service features without changing the basic functionality of the network for the establishment and the release of resources such as calls and connections.

Traffic in the signaling network is reduced by moving services closer to the customers, and the messages related to service control are handled locally. The overhead of downloading service programs is done off-line and does not impact signaling performance.

MAs enable both temporal distribution (i.e., distribution over time) and spatial distribution (i.e., distribution over different network nodes) of service logic.

MAs can be implemented in Java programming language. Additional features and mechanisms supported and envisioned in Jini programming language allow for implementation of mobile devices in practical systems.

2.1 AGENT-BASED SERVICE IMPLEMENTATION

Distributed Object Technology (DOT) provides a Distributed Processing Environment (DPE) to enable designers to create object-oriented distributed applications, which are not necessarily aware of the physical layout of the underlying network structure hidden by platform services. DOT-based specifications of DPEs, like CORBA 2.0, have been adopted by the Telecommunications Information Networking Architecture (TINA) Consortium as the basis for the distributed architecture.

Mobile Agent Technology (MAT) uses the capabilities provided by machine-independent, interpreted languages like Java to deploy a framework in which applications can roam between network nodes maintaining their execution status. MAT platforms are often based on a CORBA DPE layer that allows distributed applications to dynamically reconfigure their layout according, for instance, to processing needs. This way certain MAs may have a CORBA interface enabling them to exploit the facilities offered by the distributed objects communication infrastructure.

This framework provides service designers with additional flexibility by using CORBA object location and object interfacing facilities, and by using code migration capabilities to dynamically upgrade network nodes with new applications.

The application of DOT and MAT to the IN architecture provides benefits to the service provisioning process as shown in Figure 2.1, with maintaining the basic principle of IN related to call and service separation.

The introduction of DOT and MAT at the service design and deployment level allows for reusability for easy and rapid deployment of services, extensibility towards new and updated services, and flexibility of service design. The adoption of DOT and MAT within the Service Switching Points (SSPs) allows for services distribution among the switches with faster handling of service requests, more reliable service execution, and network scalability.

In the IN architecture, the control of the network resources is performed by the signaling plane, whereas the service creation, deployment, and provisioning is performed by the service plane. This separation allows introduction of new services and service features

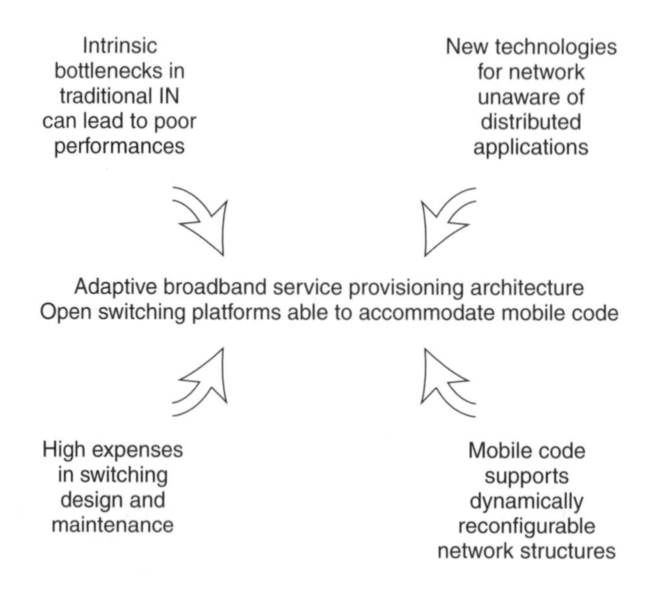

Figure 2.1 Application of DOT and MAT to the IN.

without changing the basic functionality of the network for the establishment and the release of resources such as calls and connections.

In the IN architecture, the intelligence is kept inside the core network that reduces the need to update the equipment of the Access Network (AN) representing the most widespread and expensive portion of the overall network. The IN architecture shown in Figure 2.2 comprises functional entities mapped into physical elements.

The communication between network entities is done through Signaling System No. 7 (SS7). The Intelligent Network Application Protocol (INAP) also uses SS7 for the IN

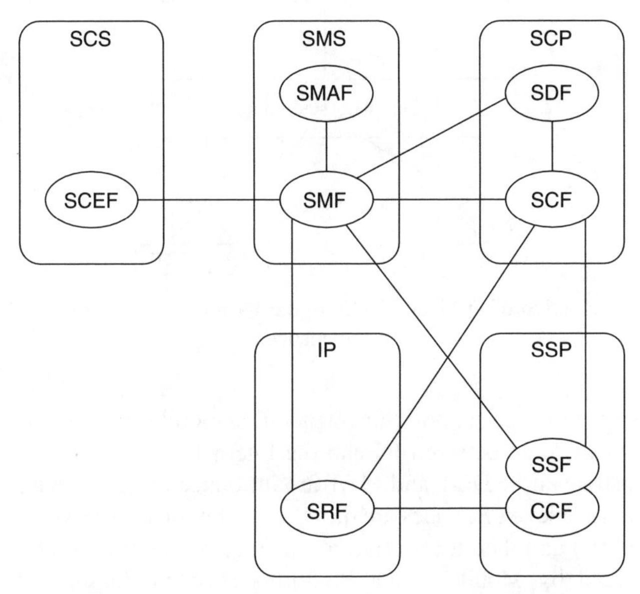

Physical entities	
SCS	Service Creation System
SMS	Service Management System
SCP	Service Control Point
IP	Intelligent Peripheral
SSP	Service Switching Point

Functional entities	
SSF	Service Switching Function
CCF	Call Control Function
SDF	Service Data Function
SCEF	Service Creation Environment Function
SRF	Specialized Resource Function
SCF	Service Control Function
SMF	Service Management Function
SMAF	Service Management Access Function

Figure 2.2 Deployment of functional entities to physical entities in the IN.

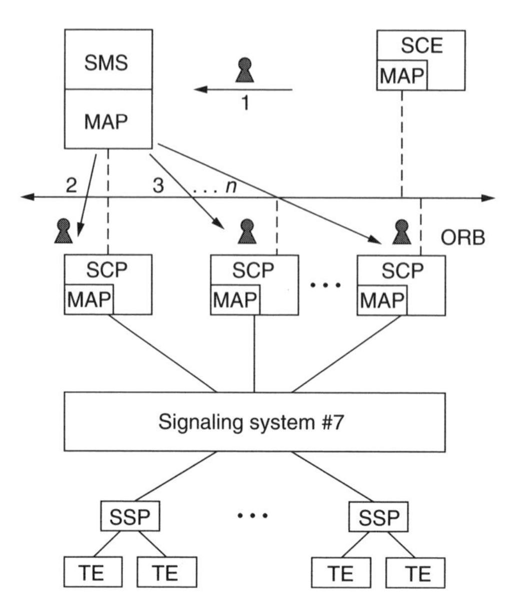

Figure 2.3 Introduction of DOT and MAT in the IN for service design, deployment, and maintenance.

messages. IN architecture can support third-generation mobile systems and has the capacity of the third-party call setup between IN and the Internet.

Figure 2.3 illustrates how DOT and MAT are introduced at the service design, deployment, and maintenance level. Services are designed as Java-based MAs in Service Creation Environments (SCEs) and then transferred to the Service Control Points (SCPs) by using capabilities provided by Mobile Agent Platforms (MAPs). In this architecture, SCPs contain CORBA and MAT in their design. Service providers benefit from a flexible service-provisioning environment by adopting object-oriented techniques for software design and by using MAT facilities to apply immediate and sophisticated policies for release distribution, update, and maintenance. Service Management System (SMS) stores and distributes services and manages the running service instances.

MAPs are introduced in the switching nodes. CORBA method invocations are used between SSPs and SCPs as an alternative to INAP as shown in Figure 2.4. The service logic (arrow 1) can be duplicated and distributed to the SCPs (arrows 2, 3, *n*), and directly to the SSPs. In this case, SS7 is only used for communication between SSPs.

This architecture with service distribution to the switches allows for faster handling of service requests, higher reliability in handing the services, scalability, and reduction of traffic in the signaling network.

Service requests are handled faster by using an agent in the switch that causes call handling, which usually does not require the establishment of a transaction with an SCP and the consequent exchange of messages in the network. Therefore, no complex protocol stacks are needed below the application part. Instead, communication between internal switch processes occurs.

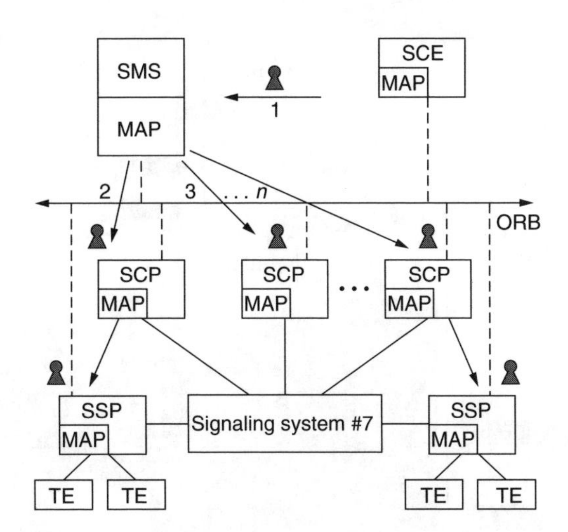

Figure 2.4 Introduction of MAPs in the IN switches.

The impact of network faults on the behavior of service is reduced since the network is accessed mainly to download the service logic. Network errors can occur during downloading Service Location Protocols (SLPs) (i.e., agent migration) or during a Remote Method Invocation (RMI) (through CORBA infrastructure). These situations can be handled by using persistent mechanisms. Most MATs offer persistent agent facilities and, for CORBA objects, the Persistent Object Service (POS) can be used. This way service performance degradation is reduced.

The problem of having centralized points is solved by distributing the service code across the network, which has a larger number of switches than SCPs. Dynamic SLP/SDT (Service Description Table) distribution allows IN services to be spread across the network to satisfy higher demand for those services. The distribution is performed dynamically when it is needed. In a distributed IN, the SLPs of the first IN calls are downloaded from the SMS to the SCP and then executed in the SCP. When the capacity of IN calls in SCP is exceeded, the SLPs are downloaded to the SSP, which must have the processing power and infrastructure to accomplish the new tasks (i.e., the SSP must also provide SCP functionality). This way the SCP can accommodate a higher number of calls and is restricted to the user interaction functionality [Broadband Special Resource Function (B-SRF) capability]. The distribution of the SLP to the attached SSPs can sustain the additional processing required per call.

Traffic in the signaling network is reduced by moving services closer to the customers, and the messages related to service control are handled locally. The overhead of downloading service programs is done off-line and does not impact signaling performance.

The distribution of services to the switches does not affect the IN basic principle of distinguishing between enriched call control (Call Control/Service Switching Functions, CCF/SSF) and service intelligence (Service Control Function, SCF). The detection of IN call attempts is still determined at call control level, and following that, an invocation of IN facilities is done by the switch. The difference is now in the communication technology

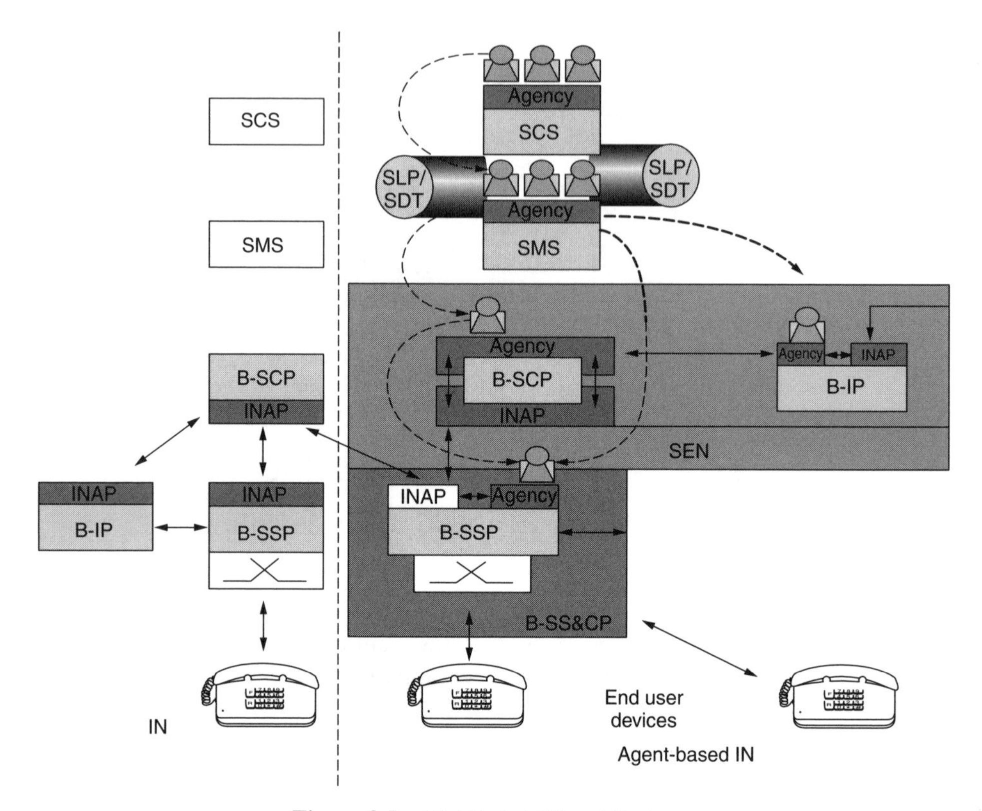

Figure 2.5 Distributed IN architecture.

between SSF and SCF, which is based on CORBA principles. Backward compatibility with traditional IN can be achieved by using IN/CORBA gateways, which allow for gradual introduction of distributed IN as advanced service islands. The distributed IN architecture is shown in Figure 2.5. In this figure, prefix B- is used with the IN functional entities to indicate the application of IN concepts to a broadband environment.

Broadband infrastructure is not a mandatory requirement and the benefits of MAT/DOT techniques to IN apply also to a narrowband architecture.

The following network elements are used in the network architecture: Service Creation System (SCS), SMS, Service Execution Node (SEN), Broadband Service Switching and Control Point (B-SS & CP), and Customer Premises Equipment. For broadband multimedia services, the terminals need to have support to access switched broadband network (e.g., ATM). They need to have specialized hardware (e.g., ATM cards) and firmware (e.g., User to Network Interface – UNI signaling stack). MAT and CORBA can be applied to network physical entities including terminals.

Services are developed and tested within SCE. The SMS provides service storage, service uploading to network elements, and service control capabilities (i.e., agent localization, alarm handling). The SEN is the physical element that joins the roles of the Broadband Service Control Point and Broadband Intelligent Peripheral. Broadband SSP has the capability to locally execute services downloaded from the network and is named B-SS & CP.

In distributed IN where CORBA can be used for message exchange, generic programming interfaces are available for developers. In this architecture, B-SCF, B-SDF, and B-SRF are implemented as CORBA-based software components allowing DPE's location transparency and direct method invocation.

There are several benefits of distributed IN architecture. The network elements can communicate in a homogeneous way. The SEN can be the contact point between the users and the network. The operator can choose a distributed, centralized service or mixed service.

Interactive Multimedia Retrieval (IMR) is an integrated multimedia service within the framework of broadband IN. Broadband Video Telephone (BVT), is a real-time, multimedia, two-party service that provides two geographically separated users with the capability of exchanging high-quality voice information, together with the transmission of high-quality video data. BVT is offered by Broadband-Integrated Services Digital Network (B-ISDN), which supports the facilities requested by the new generation of multimedia workstations.

The BVT service uses mobility management procedures to enable users to register at different (fixed) terminals. In a manner similar to the IMR and BVT services, the realization of these procedures is based on DOT and MAT.

MAs enable both temporal distribution (i.e., distribution over time) and spatial distribution (i.e., distribution over different network nodes) of service logic. In multimedia services, the porting of services usually occurs between IN elements of different types (SSPs and SCPs), whereas in mobility services, the porting of services is usually between modules of the same type (SCPs). These two approaches are not alternative and can be combined; therefore, if multimedia services are offered to mobile users, then MAT can be widespread in the IN architecture in the most effective way.

2.2 AGENT-BASED MIDDLEWARE

Terminal and user mobility are important aspects of communications systems. Laptop computers, Personal Digital Assistants (PDAs), and mobile phones are the elements of mobile office. The Agent-based Mobile Access to Multimedia Information Services (AMASE) supports agent mobility.

A mobility system that can be accessed by a user from any kind of terminal must have an appropriate device support and must be scalable, that is, the mobility system can be installed on different kinds of devices, especially mobile devices with strict resource constraints such as PDAs and mobile phones. A mobility system can be sized from a full-fledged system to a subsystem until it reaches a size and complexity that matches the constraints set by the devices involved and still provides all the required services.

The distributed AMASE Agent Environment comprises several devices and nodes, each running one instance of the stand-alone AMASE Agent Platform, which can be scaled to fit into different device types. The agent system shown in Figure 2.6 consists of two layers, the Agents System (AS) and the communication facilities. Communication

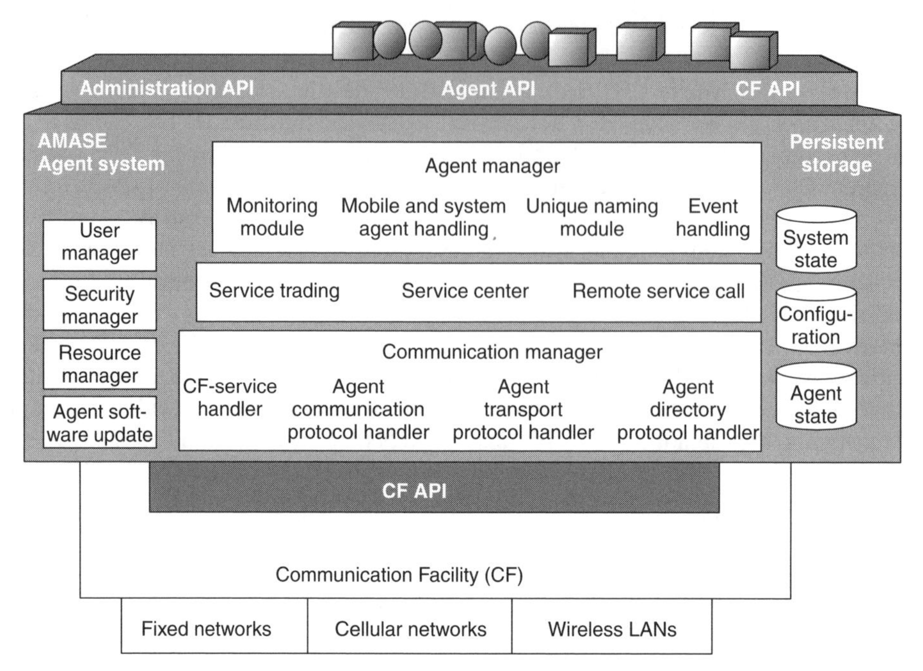

Figure 2.6 Architecture of the AMASE system.

facilities provide access to a broad range of underlying networks and handle the roaming between different kinds of networks.

The AS layer provides a runtime environment for cooperative MAs. This layer allows agents to migrate from one AS to another, to access services available in the network, and to communicate with other agents. The Service Center of the Agent System is a fundamental component for mobile agent management and user mobility and is used for locating and accessing services and agents.

The AMASE system and its supported agents are developed in Java. An agent system launcher supports loading a scaled version of the AS into a mobile device and executing it on different Java Virtual Machines (JVM). The launcher closely cooperates with a unit for agent system software update allowing for upgrading the AS's software at least at start-up or upon request. An agent launcher is used for application allowing for more convenient and browser-like launching of agent-based applications by hiding all the Java and agent system specifics.

The core of the AS is the Agent Manager (AM), allowing MAs access to the application-specific parts of the AS's functionality *via* an agent API. The communication facilities are interfaced by AS's Communication Manager (CM), and the communication facilities detect connection to available networks and their special services. The CM establishes the protocols for interagent communication, agent migration, and for accessing a Service Center and its Agent Directory (AD) *via* its protocol handlers.

The Persistent Storage area is either located in the persistent memory area of the underlying device, or on a magnetic medium. This area is needed to save agents and the agent system state and configuration.

The CM comprises user and security managers that establish a user management and allow for the enforcement of access policies. An additional resource manager provides information about device utilization, for example, memory or agent population. A component for dynamic updates of the agents' software allows for versioning and updates of agent classes.

The AM is responsible for controlling the agent population of the agent system. AM allows for launching and termination of agents and provides them with the functionality needed for migration, communication, service access, and so on. In AMASE environment, there are MAs and system agents. MAs are created by application and they can roam within the network. They are not allowed to access system resources for security reasons. Usually these agents interact with the user for an initial configuration before they are launched into the network. They allow the user to perform remote operations without a constant network connection.

MAs and system agents are supported by the AS. System agents can access system resources and become a mediator between the MAs and the system resources and the services they need to access.

The AM cooperates with the user manager and the resource manager, which permits them to assign detailed access rights to agents. Both agent types are maintained separately by the AM, which supports a clearly defined type-dependent handling, for example, in case of a shutdown. Agents are registered with the local AM, and MAs are also automatically registered with the Service Center's AD.

In Figure 2.6, the CM connects the entire agent system to the communication facilities, which connect a device to the available networks. The CM surveys preconfigured ports on sockets provided by the communication facilities to receive incoming messages. Agents can be dispatched and handled by the AM. Each CM has access either to a local or remote router provided by the agent-related directories. This router helps CM to find and address the other agent systems. The CM is responsible for converting Java objects into byte streams and is involved in synchronous communication, which requires temporal suspension of agents.

CM and communication facilities optimize communication and connection handling. The protocols consider network and device characteristics, and Quality of Service (QoS) information. Connections are physically closed during timeouts but kept open virtually. These operations that are transparent to the agents save connection costs and support disconnected operations and user mobility. The following communications mechanisms are provided by using the agent system communication manager, its protocol handlers, and the underlying communication facilities:

- asynchronous one-way agent-to-agent messages;
- synchronous two-way agent-to-agent messages based on Remote Procedure Call mechanisms;

- blackboards for local agent communication within agent systems – a blackboard is a data area where agents can leave information that may be read and removed by other agents under configurable access restrictions;
- postbox messages for specified agents; this is a message queue that belongs to a single agent and which is located at a well-known location in the network that is known to both the message senders and the postbox owner; the owner agent can only read the box contents and remove the messages, and all other agents can drop messages.

MAs are capable of migrating, which can occur at any time; thus, a mechanism is needed to determine an agent's current location. This mechanism is not necessary for asynchronous communication and communication based on blackboards and postboxes; it is inevitable for direct communication of agents. The Mobile Agent System Interoperability Facility (MASIF) specifies a Mobile Agent Facility (MAF) component MAFFinder, which is an abstract facility for mobile agent localization. MAFFinder is abstract because it does not specify how the agents are to be localized – only that a presence of such facility is required. Concepts for mobile agent localization include broadcast, forwarding, and directory service/home registry.

AMASE system introduces a Service Center based on a directory service using general mobile agent execution cycle. MAs are restricted in their size and complexity owing to the costs of agent migration. MAs use services to execute the tasks required. The agents contact a facility in the agent system that provides a naming or trading service and passes information on the location of the requested services. This Service Center in AMASE system is based on the concept introduced by the Java Agent Environment (JAE).

AMASE system introduces a ticket concept to pass information to MAs while keeping the actual migration and location information transparent. Mobile agent requesting a service from the Service Center receives a ticket shown in Figure 2.7. By calling useService (ticket), the MA uses the service provided, migrating to the respective agent system if it is not located in the same agent system. In addition to the information about home location, destination, and migration history, it is possible to store additional data in the ticket object, for instance, departure time, maximum number of connection retries, and priority information. The origin entry provides details about the creation and the starting point of the MA that is needed if the agent returns after having accomplished its task. Because of the user mobility and the disconnected operations, the originating device might be turned off and may become unreachable for the mobile agent. In this case, the permanent home entry gives an alternative address. The permanent home is an agent system at the service provider or the agent enabled home computer.

The architecture of the Service Center shown in Figure 2.8 introduces a new mechanism for localizing MAs by using the AD. Whenever a MA requests a new service or migrates to another host, its position is updated in the Service Center. The agent location is stored in the AD. This is implemented as a Lightweight Directory Access Protocol (LDAP) server, with the Service Center holding an LDAP client for accessing the AD.

In this approach, a MA's position is always known by the Service Center. The update of the agent's position is embedded in the agent migration process; a migration is not completed before the update has been executed. This way the MAs can always be tracked.

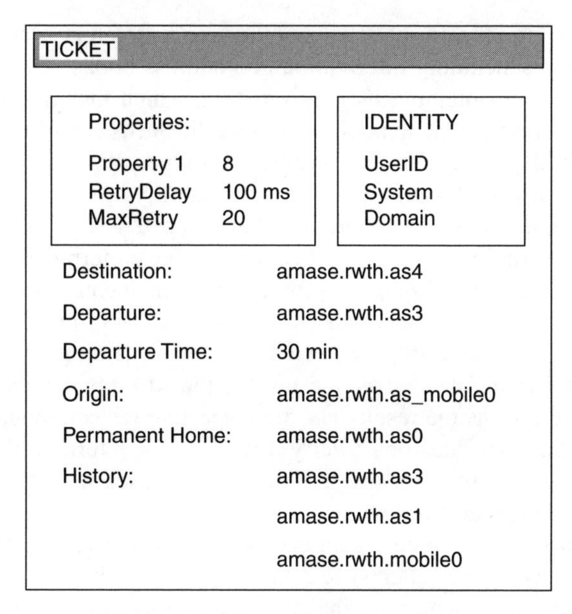

Figure 2.7 An abstract ticket object.

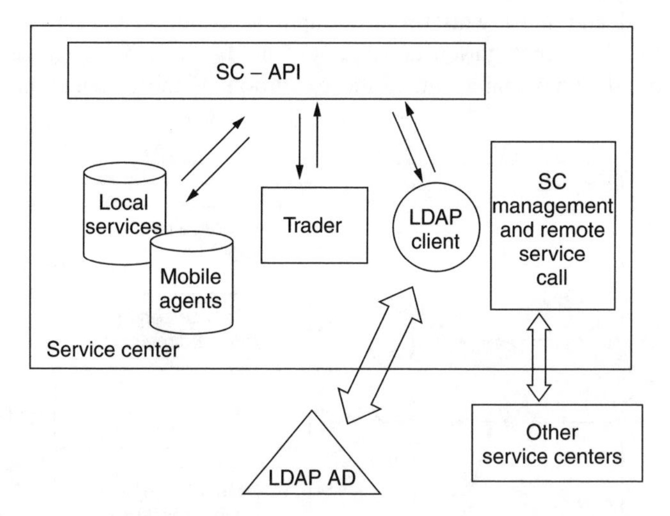

Figure 2.8 Architecture of the service center.

There are no message bursts caused by agent localization. The AD concept allows a seamless integration into the facilities required for localization services for mobile agent use.

The AMASE system allows the user to access individually configured services and data from different kinds of terminals, keeping transparent the details of the configuration and underlying mechanisms. The user profiles are in the profile directory similar to the

AD. A user profile contains information about the user's preferences and data, display and security settings, and scheduling information and address books. The profile directory is a generic database for maintaining user information, which includes application-specific data. Customized agents adapted to application-specific needs can be created on the device the user is currently deploying. The user can specify types of services to be used without having to be aware of their location or current availability.

The mobility middleware system is presented in Figure 2.9. The mobile agent, equipped with the service description and a specification of the preferred mechanism to return results, contacts the AD to localize the appropriate system agents that provide the required services. The agent obtains the ticket and migrates to the appropriate system agents and uses their services. Once the results are generated, the profile directory is used. If the user specified a type of terminal to deliver the results, the MA obtains the address from the profile directory and returns the results *via* the respective telecommunication service. On the other hand, if the user does not specify a method for returning the results, the MA decides which method to use. User and terminal profiles used with MAT create a flexible and device-independent user mobility.

The users can become temporarily unreachable when the results are available. MAs allow the users to disconnect after specifying the service. If the method specified for returning the result is an asynchronous message (e.g., e-mail, fax), no feedback is required by the MAs. On the other hand, if the agent's execution depends on the user's feedback or if the return method is selected by the user after an initial notification, the MA cannot be terminated and must wait for user input to continue execution. The AMASE system introduces the kindergarten concept for an MA, which recognized that the target user is currently unavailable, or, if the execution of the notification method failed

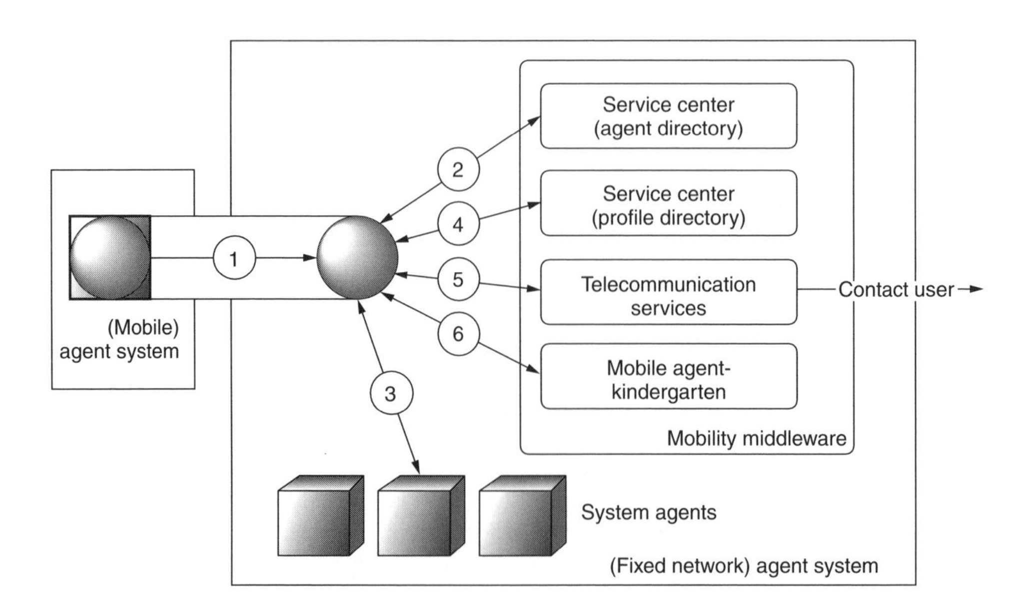

Figure 2.9 The agent-based mobility middleware.

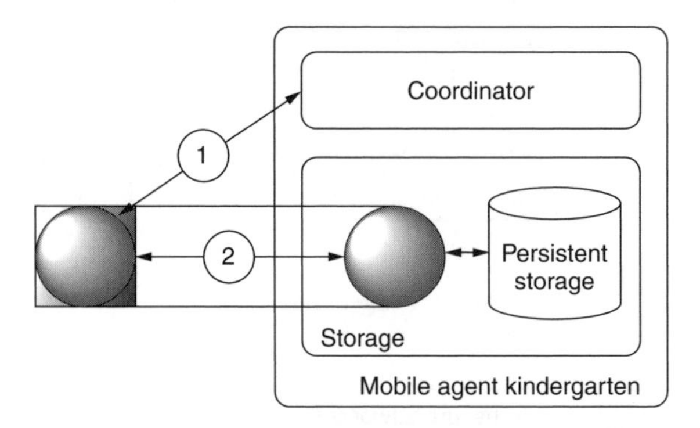

Figure 2.10 The mobile agent kindergarten concept.

or timed out, to contact a kindergarten coordinator that checks if the system having last served the MA is capable of holding this agent until the user becomes available. In this case, the agent is suspended until further notice. The agent is instructed to migrate to a host providing a kindergarten storage. This server suspends the MA and resumes it when the user reconnects. The MA can also be moved to persistent storage until being resumed, which allows for managing a large number of MAs. The kindergarten concept shown in Figure 2.10 provides a mechanism for handling MAs belonging to disconnected users and forms the basis of mobility support deploying user and terminal profiles.

2.3 MOBILE AGENT-BASED SERVICE CONFIGURATION

MAT allows for object migration and supports Virtual Home Environment (VHE) in the Universal Mobile Telecommunications System (UMTS). VHE uses MAs in service subscription and configuration.

UMTS supports QoS, the Personal Communication Support (PCS), and VHE. The VHE allows for service mobility and roaming for the user, which carries subscribed and customized services while roaming. During the registration procedure, the VHE enables the visited network to obtain the information about the user's service provider, the user's personalized service profile, and the identification about service capabilities to execute specific services.

The VHE architecture shown in Figure 2.11 can be viewed as middleware layer that hides from the user the concrete network capabilities and differences in user and provider system capabilities. Service intelligence can be located inside the network within the Service Control and Mobility Management Platform (SC & MMP) or outside the network within the Universal Service Identification Module (USIM) of the end system. Service adaptation and media conversion is needed to cope with the diversity of end systems supporting personal mobility and QoS variations of different ANs supporting terminal

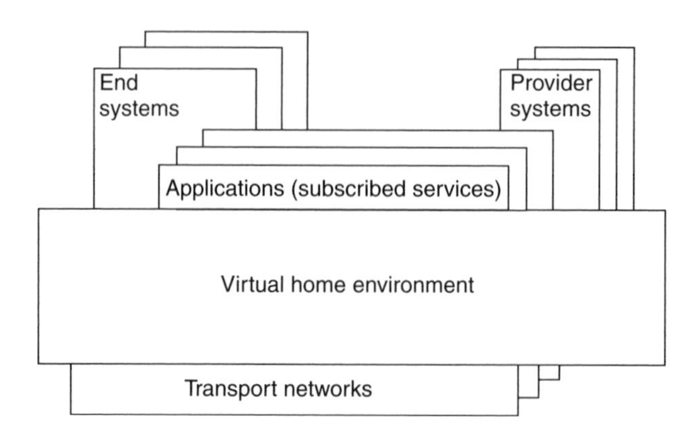

Figure 2.11 Virtual home environment.

mobility. The enhancements of service control intelligence during service execution and dynamic subscription of a new third-party services should be allowed in the system.

The UMTS environment shown in Figure 2.12 consists of a terminal, the AN, the SC & MMP, and the third-party service provider. A user registers at the terminal that presents services to the user. The user's identification and authentication is handled by the UMTS Subscriber Identity Module (USIM). The network access of the terminal is managed by the access network. Fixed or mobile terminals are linked by the AN to the SC & MMP. The SC & MMP contains service logic and is responsible for the mobility management. Third-party service provides support supplementary services. A third-party service provider has a connection to one or more SC & MMPs and does not have its own mobility management facilities.

A middleware layer is introduced in UMTS architecture in Figure 2.13. The middleware consists of Distributed Agent Environment (DAE), for example, Grasshopper, which is built on the top of DPE, for example, CORBA, and spans all potential end user systems and provider systems. The nodes provide agent environments through middleware system

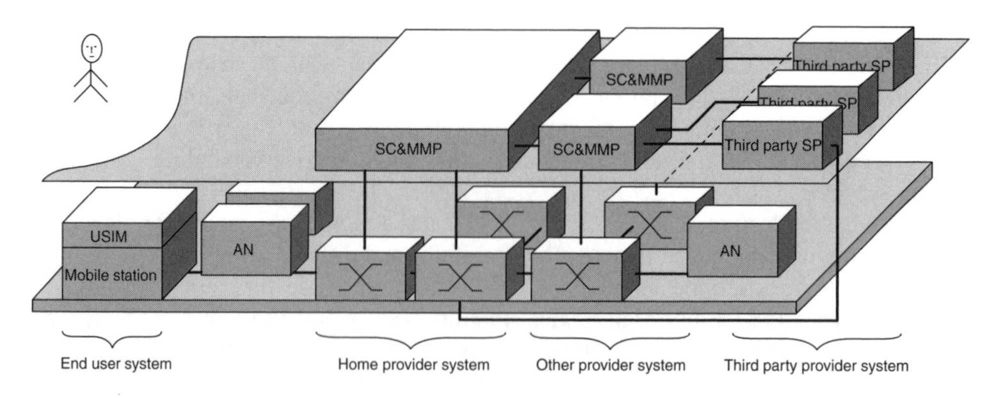

Figure 2.12 The main components of the third-generation mobile communication system.

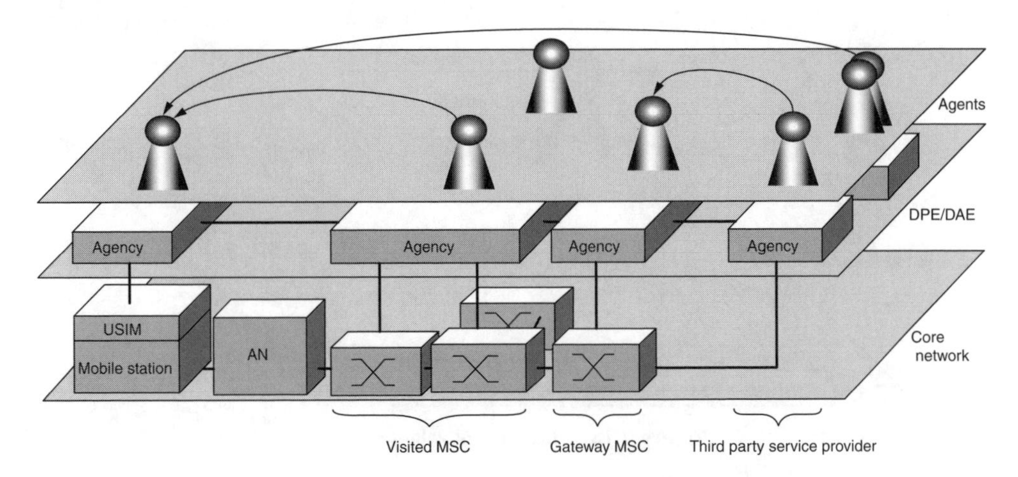

Figure 2.13 The distributed agent environment spanning across UMTS end user and provider systems.

to enable downloading and migration of MAs. MAs contain intelligence related to mobility management and service control (VHE control) and the end user application between the involved system nodes, including the Mobile Stations (MSs).

In agent-based UMTS, a VHE-agent realizes the VHE concept; a Service Agent (SA) represents a provided service; a Terminal Agent (TA) allows the terminal to inform the provider system about its capabilities; and a Provider Agent (PA) realizes a trader within the provider system, which manages all supported services (SA), that is, maintains an overview of all available services within the provider domain.

The VHE allows individually subscribed and customized services to follow their associated users wherever they roam. The VHE-agent follows the user to the domain to which the user is roaming. At every domain, the VHE-agent provides the user's subscribed services and configurations.

Agencies in the MS allow dynamic distribution of mobility management and service control intelligence to be downloaded dynamically from the MS into the (visited) provider system and from the (visited) provider system onto the MS, to be distributed within one provider system at the most appropriate location and to be distributed between different provider systems. The end systems through the USIM can take an active part in mobility management and service control.

The PA residing in every provider domain contains the knowledge of all services provided by this domain. The PA is designed as a trader in MASIF. The PA is the initial contact point of the VHE-agent after the user is roamed to a new domain. The PA is designated as a stationary agent since its task makes the migration of this agent not necessary.

The SAs are located within the provider domain, or at the third-party service provider domains, or at the user's terminal. The Converter Agents (CAs) at the provider agency are responsible for converting incoming and outgoing calls on the basis of user and terminal requirements. This allows for support of services on terminals that cannot originally

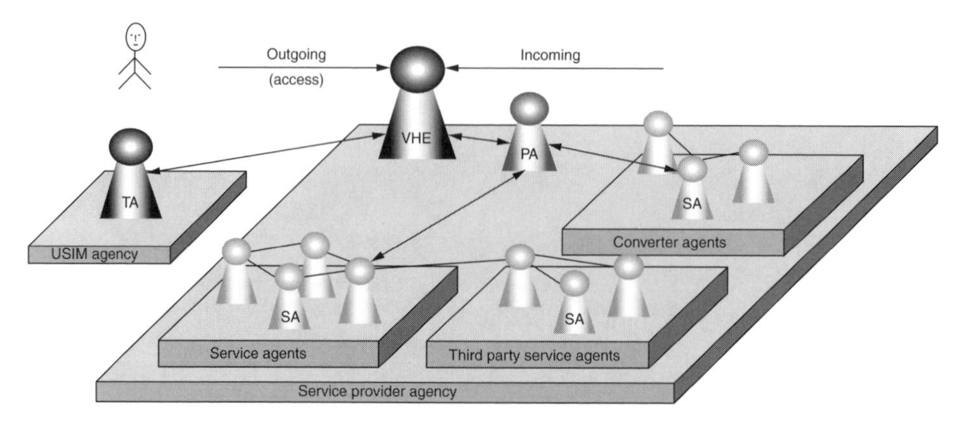

Figure 2.14 Basic agent relationships.

present the service such as reading out a fax or e-mail on a telephone. The knowledge of the terminal capabilities is maintained by the TA. Different types of agents and their communication relationships are shown in Figure 2.14.

The VHE-agent can migrate from the provider domain from which the user comes to the provider domain to which the user is roaming. Another possibility is to store a major copy of the VHE-agent within the home service provider domain. Whenever the user roams to a new provider domain, a copy of the VHE-agent migrates to this domain. The VHE-agent can also be stored on the terminal agency. The VHE-agent migrates from the terminal agency to the provider agency when the user roams to a new domain.

Dynamic subscription allows a user to subscribe to and to unsubscribe services. The subscription component presents the entire set of provided services to the user. The information can be retrieved during the registration procedure after the user roamed to a new provider, and the subscription component requests the provider to get information about provided services. The services of a new provider can also be concatenated to the service list that is stored by the VHE. The network can provide a roaming broker that can be contacted by the subscription component to get the information about service providers.

The abstract service subscription is present at the provider where the user roams. The user registers at the new provider, and the VHE-agent contacts the PA to receive a subscription interface to process the VHE request. The PA finds an SA that corresponds to the abstract service description. The PA returns a reference to the existing SA that matches with the service description. If there is no SA matching service description, the PA finds a corresponding agent at different service providers. The PA explores possibilities illustrated in Figure 2.15. The current service provider can contact other service providers, or a roaming broker can be used, or the home service provider can provide a reference to the service agent.

The User Interface Agent (UIA) is responsible for the presentation of the SA at the user's terminal. The UIA provides terminal-dependent service presentation capabilities. The same service can be represented by many UIAs for terminals with different capabilities. The VHE-agent decides which UIA download to the terminal as shown in Figure 2.16. The VHE-agent contacts the TA, which resides on the terminal agency, that

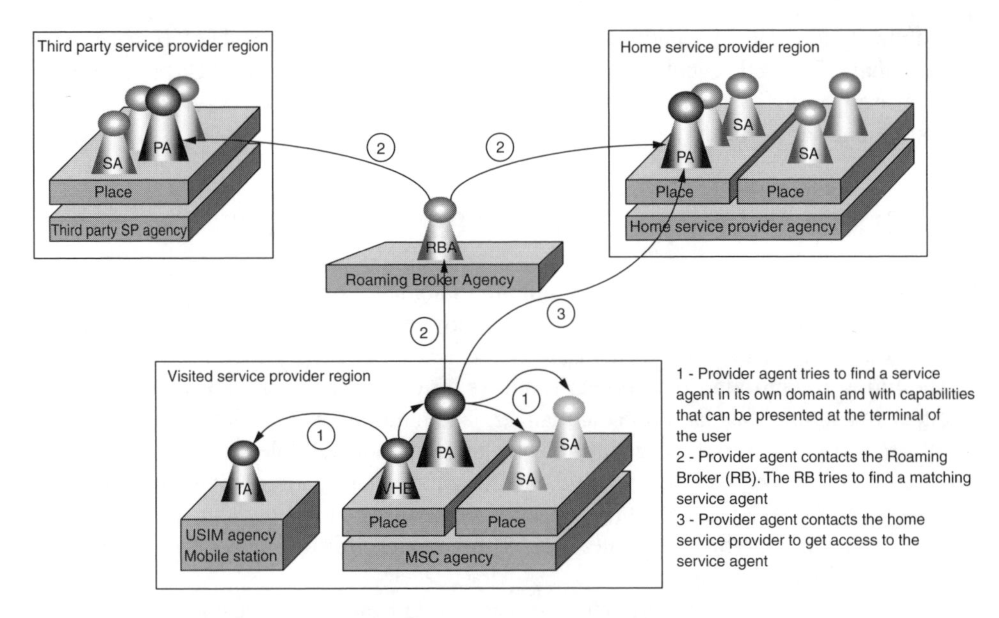

Figure 2.15 Service access strategies.

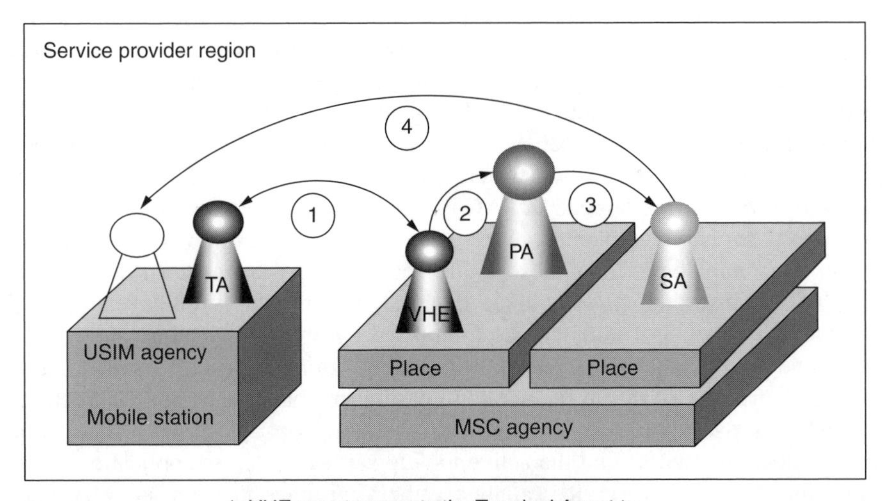

1- VHE-agent requests the Terminal Agent to get properties of the terminal. This information will be sent by the TA to the VHE-agent.
2 - The VHE-agent will request the provider agent.
3 - To find a corresponding UIA.
4 - If a UIA was found, it will be downloaded to the terminal agency.

Figure 2.16 The UIA selection procedure.

is, mobile station. This TA is device dependent and contains technical information about the terminal. The VHE-agent requests the TA to get this information. The returned values are used by the VHE-agent to find a corresponding UIA. The service subscription procedure is used to locate the UIA.

2.4 MOBILE AGENT IMPLEMENTATION

MAs can be implemented in Java programming language. Additional features and mechanisms supported and envisioned in Jini programming language allow for implementation of mobile devices in practical systems.

The Jini vision introduced by Edwards allows for devices and software services to work together in a simple, fast, and reliable manner. The requirements for devices and software specify robust software infrastructure developed to support reliable systems. The devices must be easy to use and administer, and should work instantly after being connected. The software systems must be evolvable, and software services and devices should permit their use without reconfiguration of the network. These devices form spontaneous communities within dynamic networking.

Mobile code is used in several structures to support mobile applications. In Java programming language, applets are used for small applications to be installed automatically wherever they are needed and removed when their users do not need them. In the agent paradigm, small, autonomous bits of code travel to search for desired data. The mobile code is used for performance and autonomy. Agents can provide a better performance as the code moves closer to data in the network. Agent autonomy allows the user to log off or shut down the machine, and the agent that left the originator computer can continue to run even if the originator disconnects. Java RMI allows for building various distributed systems and can be used for automatic application installation or for building agent-based systems. Mobile code in RMI is used for object-oriented networked systems and it supports evolvable implementations of remote objects and new implementations of parameter and return types. Jini uses mobile code to achieve maintenance, evolvability, and ease of administration for networked devices and services. Jini is layered atop RMI, allowing all the benefits of mobile code to be used by programs in Jini.

Jini supports spontaneously created self-healing communities of services, and it is based on the concepts of discovery, lookup, leasing, remote events, and transactions.

Jini uses discovery protocols to find the available lookup services. The Multicast Request Protocol is used to find the active lookup services after an application or service becomes initiated. Lookup services announce their existence in the system by using the Multicast Announcement Protocol. An application or service talks to the lookup service by using the Unicast Discovery Protocol.

The lookup service is a process that has semantic information about available services. The service items have proxy objects and attributes describing these services.

Jini concept of leasing allows the resource to be loaned to a customer for a fixed period of time rather than granting access to a resource for an unlimited amount of time. This ensures that the communities of services are stable, self-healing, and resilient to failures, errors, and crashes.

Jini uses remote events to allow services to notify each other about the changes in their state. These are messages sent as asynchronous notifications directly to a software component and handled outside the normal flow of control of the component.

Computations involving multiple services reach safe and known state by using transactions in Jini. Transactions provide atomicity, consistency, isolation, and durability to data manipulations. All the operations under transactions are executed as an atomic operation. Transactions ensure state consistency after completion. Transactions are isolated during execution; they do not affect one another until completion. Transaction durability makes the changes permanent.

2.5 SUMMARY

MAT uses the capabilities provided by machine-independent, interpreted languages like Java to deploy a framework in which applications can roam between network nodes maintaining their execution status. MAT platforms are often based on a CORBA DPE layer, which allows distributed applications to dynamically reconfigure their layout according, for instance, to processing needs. This way certain MAs may have a CORBA interface enabling them to exploit the facilities offered by the distributed objects communication infrastructure.

The introduction of DOT and MAT at the service design and deployment level allows for reusability for easy and rapid deployment of services, extensibility towards new and updated services, and flexibility of service design.

The AS layer provides a runtime environment for cooperative MAs. This layer allows agents to migrate from one AS to another, to access services available in the network, and to communicate with other agents. The Service Center of Agent System is a fundamental component for mobile agent management and user mobility and is used for locating and accessing services and agents.

MAT allows for object migration and supports VHE in the UMTS. VHE uses MAs in service subscription and configuration.

The UMTS environment consists of a terminal, the AN, the SC & MMP, and the third-party service provider.

In agent-based UMTS, a VHE-agent realizes the VHE concept; an SA represents a provided service; a TA allows the terminal to inform the provider system about its capabilities; and a PA realizes a trader within the provider system, which manages all supported services (SA), that is, maintains an overview of all available services within the provider domain.

PROBLEMS TO CHAPTER 2

Mobile agent-based service implementation, middleware, and configuration

Learning objectives

After completing this chapter, you are able to

- demonstrate an understanding of distributed object technology;
- discuss what is meant by intelligent agents;

- demonstrate an understanding of agent-based service implementation;
- explain how to handle service requests;
- explain temporal and spatial distribution of service logic;
- discuss multimedia services;
- demonstrate an understanding of a mobility system;
- explain what agent system, agent manager, and communication manager are;
- explain mobility middleware system;
- demonstrate an understanding of the Universal Mobile Telecommunications System (UMTS);
- discuss what an agent-based UMTS is;
- demonstrate an understanding of the VHE concept;
- explain how the VHE-agent migrates in the system;
- explain dynamic subscription of services;
- demonstrate an understanding of mobile agent implementation in Java programming language;
- demonstrate an understanding of mobile agent implementation in Jini programming language.

Practice problems

2.1: What is the role of DPE in DOT?
2.2: What are the functions of Intelligent and Mobile Agents?
2.3: What distribution of service logic is enabled by Mobile Agents?
2.4: What are the requirements for a mobility system?
2.5: What is the role of the Agent System layer?
2.6: Where is the Persistent Storage located and why is it needed?
2.7: What is supported by the UMTS?
2.8: What are the elements of the UMTS environment?
2.9: How is the VHE concept realized in agent-based UMTS?
2.10: What is the role of dynamic subscription?
2.11: What is the role of the User Interface Agent (UIA)?
2.12: What are applets used for in Java programming language?
2.13: What is the concept of Jini programming language?

Practice solutions

2.1: DOT provides a DPE to enable designers to create object-oriented distributed applications, which are not necessarily aware of the physical layout of the underlying network structure hidden by platform services.
2.2: Intelligent Agents have the ability to learn and react. MAs can migrate between different hosts, execute certain tasks, and collaborate with other agents.
2.3: MAs enable both temporal distribution (i.e., distribution over time) and spatial distribution (i.e., distribution over different network nodes) of service logic.
2.4: Mobility system that can be accessed by a user from any kind of terminal must have an appropriate device support and must be scalable, that is, the mobility system

can be installed on different kinds of devices, especially mobile devices with strict resource constraints such as PDAs and mobile phones. A mobility system can be sized from a full-fledged system to a subsystem until it reaches a size and complexity that matches the constraints set by the devices involved and still provides all the required services.

2.5: The AS layer provides a runtime environment for cooperative MAs. This layer allows agents to migrate from one AS to another, to access services available in the network, and to communicate with other agents. The Service Center of AS is a fundamental component for mobile agent management and user mobility, and it is used for locating and accessing services and agents.

2.6: The Persistent Storage area is either located in the persistent memory area of the underlying device or on a magnetic medium. This area is needed to save agents and the agent system state and configuration.

2.7: UMTS supports QoS, the PCS, and VHE.

2.8: The UMTS environment consists of a terminal, the AN, the SC & MMP, and the third-party service provider.

2.9: In agent-based UMTS, a VHE-agent realizes the VHE concept; an SA represents a provided service; a TA allows the terminal to inform the provider system about its capabilities; and a PA realizes a trader within the provider system, which manages all supported services (SA), that is, maintains an overview of all available services within the provider domain.

2.10: Dynamic subscription allows a user to subscribe to and to unsubscribe services. The subscription component presents the entire set of provided services to the user.

2.11: The UIA is responsible for the presentation of the SA at the user's terminal. The UIA provides terminal-dependent service presentation capabilities. The same service can be represented by many UIAs for terminals with different capabilities.

2.12: In Java programming language, applets are used for small applications to be installed automatically wherever they are needed, and removed when their users do not need them.

2.13: Jini supports spontaneously created self-healing communities of services and is based on the concepts of discovery, lookup, leasing, remote events, and transactions.

3

Wireless local area networks

Virtual LANs provide support for workgroups that share the same servers and other resources over the network. A flexible broadcast scope for workgroups is based on Layer 3 (network). This solution uses multicast addressing, mobility support, and the Dynamic Host Configuration Protocol (DHCP) for the IP. The hosts in the network are connected to routers *via* point-to-point connections. The features used are included in the IPv6 (Internet Protocol version 6) protocol stacks. Security can be achieved by using authentication and encryption mechanisms for the IP. Flexible broadcast can be achieved through enhancements to the IPv6 protocol stack and a DHCP extension for workgroups.

Orthogonal Frequency Division Multiplex (OFDM) is based on a mathematical concept called *Fast Fourier Transform* (FFT), which allows individual channels to maintain their orthogonality or distance to adjacent channels. This technique allows data symbols to be reliably extracted and multiple subchannels to overlap in the frequency domain for increased spectral efficiency. The IEEE 802.11 standards group chose OFDM modulation for wireless LANs operating at bit rates up to 54 Mb s^{-1} at 5 GHz.

Wideband Code Division Multiple Access (WCDMA) uses 5 MHz channels and supports circuit and packet data access at 384 kb s^{-1} nominal data rates for macrocellular wireless access. WCDMA provides simultaneous voice and data services. WCDMA is the radio interface technology for Universal Mobile Telecommunications System (UMTS) networks.

Dynamic Packet Assignment (DPA) is based on properties of an OFDM physical layer. DPA reassigns transmission resources on a packet-by-packet basis using high-speed receiver measurements. OFDM has orthogonal subchannels well defined in time–frequency grids, and has the ability to rapidly measure interference or path loss parameters in parallel on all candidate channels, either directly or on the basis of pilot tones.

3.1 VIRTUAL LANs

Virtual LANs provide support for workgroups. A LAN consists of one or more LAN segments, and hosts on the same LAN segment can communicate directly through Layer 2 (link layer) without a router between them. These hosts share the same Layer 3 (network layer) subnet address, and communication between the hosts of one LAN segment remains in this segment. Thus Layer 3 (network layer) subnet address forms a broadcast scope that contains all hosts on the LAN segment.

The workgroups are groups of hosts sharing the same servers and other resources over the network. The hosts of a workgroup are attached to the same LAN segment, and broadcasting can be used for server detection, name resolution, and name reservation.

In a traditional LAN the broadcast scope is limited to one LAN segment. Switched LANs use a switch infrastructure to connect several LAN segments over high-speed backbones. Switched LANs share the Layer 3 (network layer) subnet address, but offer an increased performance compared to traditional LANs, since not all hosts of a switched LAN have to share the bandwidth of the same LAN segment. LAN segments connected over backbones allow for distribution of hosts over larger areas than that covered by a single LAN segment.

Traditional switched LANs require a separate switch infrastructure for each workgroup in the environment with several different workgroups using different LAN segments. Virtual LANs are switched LANs using software configurable switch infrastructure. This allows for creating several different broadcast scopes over the same switch infrastructure and for easily changing the workgroup membership of individual LAN segments.

The disadvantage of virtual LANs is that a switch infrastructure is needed and administration includes Layers 2 and 3 (link and network). A desirable solution involves only Layer 3 (network) and does not require special hardware.

Kurz *et al.* propose a flexible broadcast scope for workgroups based on Layer 3 (network). This solution uses multicast addressing, mobility support, and the DHCP for the IP. The hosts in the network are connected to routers *via* point-to-point connections. The features used are included in the IPv6 protocol stacks. Security can be achieved by using authentication and encryption mechanisms for the IP. Flexible broadcast can be achieved through enhancements to the IPv6 protocol stack and a DHCP extension for workgroups.

In IPv6, a special address range is reserved for multicast addresses for each scope, and a multicast is received only by those hosts in this scope that are configured to listen to this specific multicast address. To address all hosts in a certain scope with a multicast, the multicast must be made to the predefined all-nodes address, to which all hosts must listen. When existing software using IPv4 (Internet Protocol version 4) is migrated to IPv6, the IPv4 broadcasts are changed to multicasts to the all-nodes address, as this is the simplest way to maintain the complete functionality of the software.

IPv6 multicasting can be used to form the broadcast scope of a workgroup. The workgroup is the multicast group, whose hosts listen to the same multicast address, the workgroup address. A host can listen to several multicast addresses at the same time and can be a member of several workgroups.

Multicasting exists optionally for IPv4 and is limited by a maximum of hops. The multicast in IPv6 is limited by its scope, which is the address range.

In a virtual LAN, the workgroup membership of a host is determined by configuration of the switches. Kurz *et al.* propose that a host has to determine its workgroups and their corresponding multicast addresses. Different workgroups are separated in Layer 3 (network) since each host has the possibility to address a specified subset of hosts of the network using multicasting. All hosts can be connected directly to the routers, and the members of different workgroups can share the same LAN segment.

The administration of the workgroups is designed by storing the information about hosts and their workgroups in a central database in a DHCP server. The information is distributed by using the Dynamic Host Configuration Protocol version 6 (DHCPv6).

3.1.1 Workgroup management

In a workgroup address configuration, the host sends a DHCP Request with a Workgroup Address Extension to the DHCP Server. The DHCP Server replies with a Workgroup Address Extension containing all workgroup addresses assigned to this host. After receiving the workgroup addresses, the host sends the Internet Control Message Protocol version 6 (ICMPv6) Group Membership Report to each of its workgroup addresses to inform the multicast routers about its new membership in these multicast groups.

After learning its workgroup addresses, the host has to configure its interfaces to listen to these multicast addresses. The host has to change all outgoing multicasts to the all-nodes address (which are equivalent to IPv4 broadcasts) to multicast to the workgroup address of the host. This can be done by changing the IPv6 stack to intercept all outgoing multicasts to the all-nodes address and to change this address to the workgroup addresses of the host. If the host is a member of several workgroups, the multicast has to be sent to all workgroup addresses of the host.

The purpose of DHCP is to provide hosts with addresses and other configuration information. DHCP delivers the configuration data in extensions that are embedded in request, reply, or reconfigure messages. The request message is used by the client to request configuration data from the server, and the reply message is used by the server to return the requested information to the client. If there is a change in the DHCP database, the server uses the reconfigure message to notify the client about the change and to start the new request reply cycle.

Kurz *et al.* introduce a DHCP Workgroup Address Extension to deliver workgroup addresses to the host. In a DHCP Request the client must set the workgroup count to zero, must not specify any workgroup addresses, and must specify its node name. In a DHCP Reply the server must set the workgroup count to the number of workgroup addresses existing for this client, include all workgroup addresses existing for this client, and use the client's node name. In a DHCP Reconfigure the server must set the workgroup count to zero, must not specify any workgroup addresses, and must use the client's node name.

Mobile hosts can be the members of workgroups. The Internet draft Mobility Support in IPv6 proposes that a mobile host attached to a network segment other than its home segment continues to keep its home address on the home segment and forms a global care-of address for its new location. The binding update options included in IPv6 packets are used to inform correspondent hosts as well as the home agent, a router that is on the same segment as the home address of the mobile host, about its new care-of address. After the home agent is informed about the new care-of address of the mobile host, the home agent receives packets on the home segment addressed to the mobile host and tunnels them to the care-of address of the mobile host.

Kurz *et al.* propose enhancements to the Internet draft Mobility Support in IPv6 for a mobile workgroup member to send or receive multicast packets from its home network and to participate in the multicast traffic of its group. If a mobile host leaves the scope of a multicast group it joined, the home agent must forward packets sent to the home address of the mobile host and also all packets sent to the concerned multicast address. The mobile host has to be able to send packets to the multicast address of its workgroup, even though it is outside the scope of this address. This can only be done by tunneling the packets to a host inside the scope of the multicast address and resending them from that host. Since the home agent is on the segment associated with the home address of the mobile host, the task of resending multicasts of a mobile host can also be taken over by the home agent.

The Internet draft Mobility Support in IPv6 proposes a binding update option, which is used to notify the home agent and other hosts about a new care-of address of a mobile host. The original home link local address of the mobile host has to be specified in the source address field in the IP header of the packet containing the binding update option. It can also be specified in the home link local address field in the binding update option, but a multicast address cannot be specified this way. Kurz *et al.* introduce an optional field for a multicast address in the binding update option to inform the home agent about workgroup addresses to which the mobile host listens. A field for the workgroup address is used to indicate that there is a multicast group address specified in the option.

3.1.2 Multicast groups

A mobile host that left the scope of one of its multicast groups sends a binding update option to its home agent to inform it about the new care-of address. A mobile host has to specify its multicast group address in the binding update option. If the mobile host is a member of several multicast groups, it has to send a binding update option for each of its multicast groups.

A home agent notified by a binding update option about a multicast address for a mobile host must join this multicast group and handle packets with this multicast address in the destination address field in the same way as the packets with the home address of the mobile node in this field. The mobile host must treat a received encapsulated multicast packet in the same way as the packet received directly. The mobile host must not send a binding update option to the address specified in the source address field of an encapsulated multicast packet.

When sending a multicast packet to its multicast group, the mobile host has to use its home address in the source address field of the multicast packet and tunnel this packet to its home agent. When a home agent receives an encapsulated multicast packet in which the source address field is the same as the home address of a mobile host served by it, the home agent has to act like a router, receiving this multicast packet from the home segment of the mobile host and additionally forwarding it to the home segment of the mobile host.

This way of providing mobile workgroup members with the possibility to leave the scope of the multicast address has a drawback that it may not scale well in the case of broadcast intensive workgroup protocol stacks, since all the broadcasting traffic, which was intended to remain in the limited area, has to be forwarded to the mobile node. If many workgroup members use the possibility of global mobility, there is a risk of overloading the Internet with workgroup broadcasting traffic.

Virtual LANs enhance the flexibility of the available software without requiring any changes to the software. The software adapted in the new IPv6 address space in the future can be changed to use the all-nodes multicast address instead of IPv4 broadcast. When using IPv6 multicasting, no special Virtual LAN switches and protocols are required, and only small enhancements to IPv6 and DHCP are necessary. This solution can offer a viable software alternative to Virtual LANs when faster routers are available.

3.2 WIDEBAND WIRELESS LOCAL ACCESS

3.2.1 Wideband wireless data access based on OFDM and dynamic packet assignment

OFDM has been shown to be effective for digital audio and digital video broadcasting at multimegabit rates. The IEEE 802.11 standards group chose OFDM modulation for Wireless LANs operating at bit rates up to 54 Mb s^{-1} at 5 GHz.

OFDM has been widely used in broadcast systems, for example, for Digital Audio Broadcasting (DAB) and for Digital Video Broadcasting (DVB). OFDM was selected for these systems primarily because of its high spectral efficiency and multipath tolerance. OFDM transmits data as a set of parallel low bandwidth (from 100 Hz to 50 kHz) carriers. The frequency spacing between the carriers is a reciprocal of the useful symbol period. The resulting carriers are orthogonal to each other, provided correct time windowing is used at the receiver. The carriers are independent of each other even though their spectra overlap. OFDM can be easily generated using an Inverse Fast Fourier Transform (IFFT) and it can be received using an FFT. High data rate systems are achieved by using a large number of carriers (i.e., 2000–8000 as used in DVB). OFDM allows for a high spectral efficiency as the carrier power, and modulation scheme can be individually controlled for each carrier.

Chuang and Sollenberger proposed OFDM modulation combined with DPA, with wideband 5-MHz channels for high-speed packet data wireless access in macrocellular and microcellular environments, supporting bit rates ranging from 2 to 10 Mb s^{-1}. OFDM can largely eliminate the effects of intersymbol interference for high-speed transmission rates in very dispersive environments. OFDM supports interference suppression and space–time coding to enhance efficiency. DPA supports spectrum efficiency and high-rate data access.

Chuang and Sollenberger proposed DPA based on properties of an OFDM physical layer. DPA reassigns transmission resources on a packet-by-packet basis using high-speed receiver measurements. OFDM has orthogonal subchannels well defined in time–frequency grids and has the ability to rapidly measure interference or path loss parameters in parallel on all candidate channels, either directly or on the basis of pilot tones.

The protocol for a downlink comprises of four steps:

1. A packet page from a base station to a terminal.
2. Rapid measurements of resource usage by a terminal using the parallelism of an OFDM receiver.
3. A short report from the terminal to the base station of the potential transmission quality associated with each radio resource.
4. Selection of resources by the base and transmission of the data.

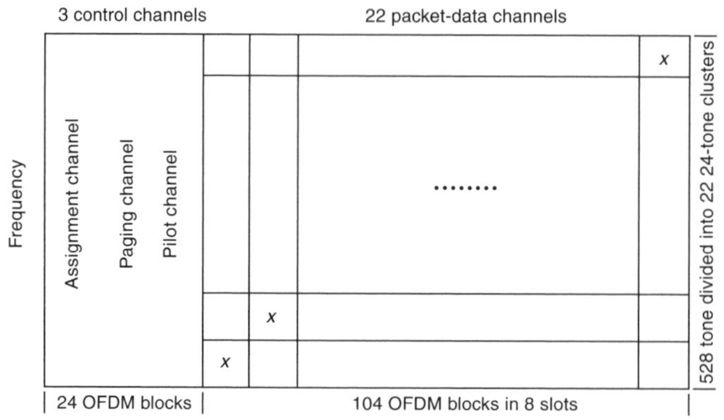

	3 control channels			22 packet-data channels	
Frequency	Assignment channel	Paging channel	Pilot channel	....	528 tone divided into 22 24-tone clusters
	24 OFDM blocks			104 OFDM blocks in 8 slots	

Figure 3.1 Division of radio resources in time and frequency domains to allow DPA for high peak-rate data services.

The frame structures of adjacent Base Stations (BSs) are staggered in time; the neighboring BSs sequentially perform the four different DPA functions with a predetermined rotational schedule. This avoids collision of channel assignments. This protocol provides a basis for admission control and bit rate adaptation based on measured signal quality.

Figure 3.1 shows radio resources allocation scheme in which 528 subchannels, each of 4.224 MHz, are organized into 22 clusters of 24 subchannels of 192 kHz each in frequency and 8 time slots of 13 OFDM blocks each within a 20 ms frame of 128 blocks. This allows flexibility in channel assignment while providing 24 blocks of control overhead to perform the DPA procedures. Each tone cluster contains 22 individual modulation tones plus two guard tones. There are 13 OFDM blocks in each traffic slot and two blocks are used as overhead – a leading block for synchronization and a trailing block as guard time for separating consecutive time slots. A radio resource is associated with a frequency hopping pattern in which the packets are transmitted using eight different tone clusters in each of the eight traffic slots. Coding across eight traffic slots for user data exploits frequency diversity, which gives sufficient coding gain for performance enhancement in the fading channel. This arrangement supports 22 resources in frequency that can be assigned by DPA. Considering overhead for OFDM block guard time, synchronization, slot separation, and DPA control, a peak data rate of 2.1296 ($3.3792 \times 22/24 \times 11/13 \times 104/128$) Mb s^{-1} is available for packet data services using all 22 radio resources, each of 96.8 kb s^{-1}.

Frame structure is shown in Figure 3.2 for downlink DPA. The uplink structure is similar but the control functions are slightly different. In each frame the control channels for both the uplink and downlink jointly perform the four DPA procedures sequentially with a predetermined staggered schedule among adjacent BSs. The control channel overhead is included to allow three sectors to perform DPA at different time periods. This allows interference reduction and additional Signal to Interference Ratio (SIR) enhancement for the control information. Spectrum reuse is achieved for traffic channels through interference avoidance using DPA to avoid slots causing potential interference. The frame structure

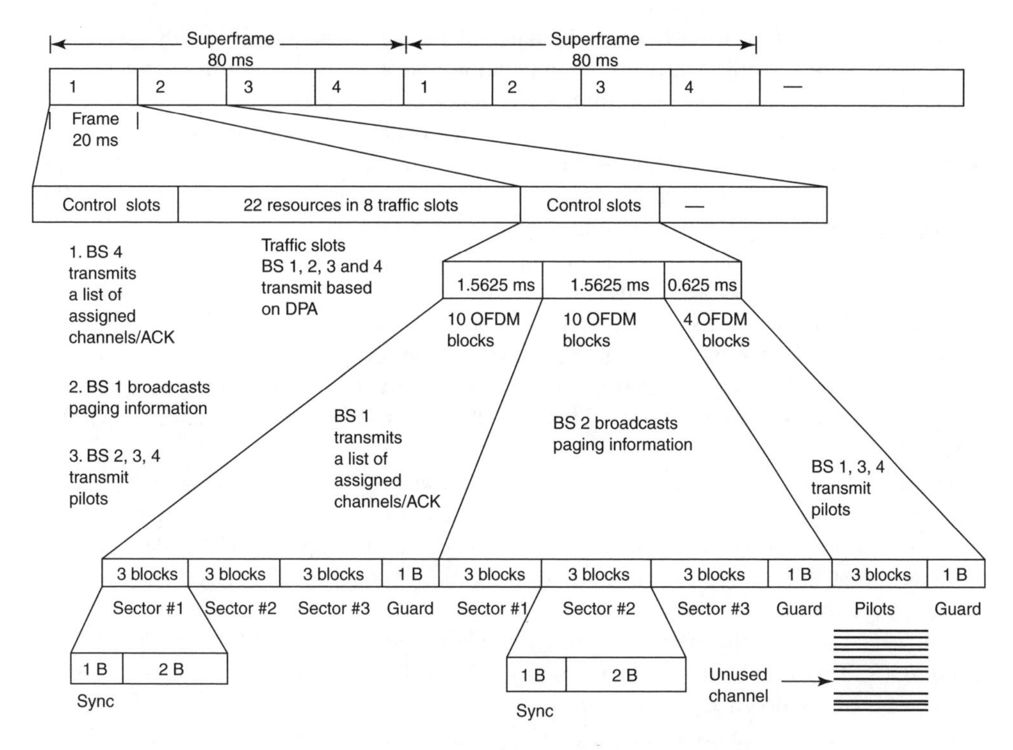

Figure 3.2 Frame structure for downlink DPA.

permits SIR estimation on all unused traffic slots. The desired signal is estimated by the received signal strength from the two OFDM blocks used for paging. The interference is estimated by measuring three blocks of received pilot signals. The pilot channels are generated by mapping all the radio resources currently in use onto corresponding pilot subchannels, thus providing an interference map without monitoring the actual traffic subchannels. The OFDM scheme handles many subchannels in parallel, which allows for fast SIR estimation. The measurement errors are reduced through significant diversity effects with 528 available subchannels to map 22 resources over three OFDM blocks. The estimated SIR is compared against an admission threshold (for instance, 10 dB), and channel occupancy can be controlled to achieve good Quality of Service (QoS) for the admitted users.

3.2.2 Wireless services support in local multipoint distribution systems

Several systems support broadband wireless communications and mobile user access. These are the Multichannel Multipoint Distribution System (MMDS) and the Local Multipoint Distribution System (LMDS), also called Local Multipoint Communication System (LMCS) or Microwave Video Distribution System (MVDS).

The MMDS systems work at frequencies lower than 5 GHz in large coverage areas with cell radius of up to 40 km. MMDS systems can be used for transmission of video and broadcast services in rural areas. Because of the large cell size, MMDS systems do not perform well for bidirectional communication that integrates a return channel.

The LMDS systems work with higher frequencies where a larger frequency spectrum is available than that in the MMDS systems. The coverage for LMDS systems involves smaller cells of up to 5 km radius and requires repeaters to be placed in a Line Of Sight (LOS) configuration. This local coverage with a large available bandwidth makes LMDS systems suitable for interactive multimedia services distribution.

Broadband wireless access is based on the Two-Layer Network (TLN) concept in which subscribers are grouped into microcells, which are embedded into a macrocell. The microcells coverage uses local repeaters operating at 5.8 GHz fed by a BS through 40 GHz links. OFDM modulation is used to allow the reception with plug-free receivers located inside the buildings. A 40 GHz band fixed receiver provides a rooftop antenna in LOS with the transmitting antenna. This LMDS system provides an integrated wireless return channel.

The LMDS architecture uses co-sited BS equipment. The indoor digital equipment connects to the network infrastructure, and the outdoor microwave equipment mounted on the rooftop is housed at the same location. The Radio Frequency (RF) planning uses multiple sector microwave systems, where the cell site coverage is divided into 4, 8, 12, 16, or 24 sectors.

The user accesses the network through Hybrid Fiber Radio (HFR), Radio To The Building (RTTB) and Radio To The Curb (RTTC). In HFR, a Radio Frequency Unit (RFU) carries out signal down conversion from RF frequency to the intermediate frequency. The signal feeds the Radio Termination (RT) of each user through a bus link. In RTTB architecture the signal feeds the user Network Termination (NT) through point-to-point cable links. In RTTC the RFU is placed in a common outdoor unit and is shared among several buildings.

In high-population cities, LMDS systems can be used as LOS propagation channels at high frequencies. LOS operation is inherently inflexible even for low mobility services. On the other hand, the available bandwidth for LMDS frequencies exceeds 1 GHz, making it a very desirable transmission method. The frequency bands assigned to MMDS and LMDS are included in the frequency bands allocated for fixed services. The exception is the 40.5–42.5-GHz band allocated for MVDS systems. The 28-GHz channel is not generally open in several countries. This is why the 40-GHz technology is considered. However, the baseband system is designed to be compatible with interchangeable RF system (5/17/28/40 GHz).

LMDS is a stand-alone system providing wireless multimedia and Internet services, and it can be used as the support infrastructure for other wireless multimedia services, for example, UMTS, wireless LAN, and Broadband Radio Access Network (BRAN), which provide a high-speed digital connection to the user.

Sukuvaara *et al.* proposed a two-layer 40-GHz LMDS system providing wireless interactive cellular television and multimedia network. The first layer, a macrocell, uses 40-GHz wireless connection between the BS and the sub–base station, which can be a frequency and/or protocol conversion point called a *local repeater*. The second layer, a microcell, operates at 5.8 GHz. The user can connect a multimedia PC (Personal Computer) to a local repeater access point at 5.8 GHz or directly to the BS at 40 GHz. The

5.8 GHz connection can be used cost effectively within cities and high-density population areas, and the 40 GHz connection can be used in rural areas. The macrocell size can be up to 5 km. The microcell size is from 50 to 500 meters depending on services and location. A 40-GHz transceiver unit serves dozens of microcell users. The microcell architecture prevents LOS indoor propagation, supports nomadic terminals, and is cost effective.

3.2.3 Media Access Control (MAC) protocols for wideband wireless local access

Wireless LANs provide wideband wireless local access and offer intercommunication capabilities to mobile applications. This technology is supported by 802.11 standard developed by the IEEE 802 LAN standards organization. Wireless LANs are also provided by High Performance Radio LAN (HIPERLAN) Type 1 defined by the European Telecommunications Standards Institute (ETSI) RES-10 Group.

IEEE 802.11 uses data rates up to 11 Mb s^{-1} and defines two network topologies. The infrastructure-based topology allows Mobile Terminals (MTs) to communicate with the backbone network through an access point. In *ad hoc* topology, MTs communicate with each other without connectivity to the wired backbone network. HIPERLAN uses data rate 23.5 Mb s^{-1} and the *ad hoc* topology.

QoS guarantees are achieved through infrastructure topology, and a priority scheme in the Point Coordination Function (PCF) in the IEEE 802.11. HIPERLAN defines a channel access priority scheme based on the lifetime of packets to achieve QoS.

Wireless Asynchronous Transfer Mode (WATM) standardization involves Wireless ATM Group (WAG) of the ATM Forum and the BRAN project of ETSI. These efforts involve developing a technology for wideband wireless local access that includes ATM features in the radio interface, thus combining support of user mobility with statistical multiplexing and QoS guarantee provided by wired ATM networks. The goal is to reduce complexity of interworking between the wireless access network and the wired ATM backbone and to attain a higher level of integration.

3.2.4 IEEE 802.11

The IEEE 802.11 MAC (Media Access Control) protocol provides asynchronous and synchronous (contention-free) services, which are provided on top of physical layers and for different data rates. The asynchronous service is mandatory, and the synchronous service is optional.

The asynchronous service is provided by the Distributed Coordination Function (DCF), which implements the basic access method of the IEEE 802.11 MAC protocol also known as Carrier Sense Multiple Access with Collision Avoidance (CSMA/CA) protocol. The implementation of DCF is mandatory.

Contention-free service is provided by the PCF, which implements a polling access method. A point coordinator cyclically polls wireless stations, allowing them to transmit. The PCF relies on the asynchronous service provided by the DCF. The implementation of the PCF is not mandatory.

Basic access mechanism illustrated in Figure 3.3 explains that in DCF a station must sense the medium before initiating transmission of a packet. If the medium is sensed to

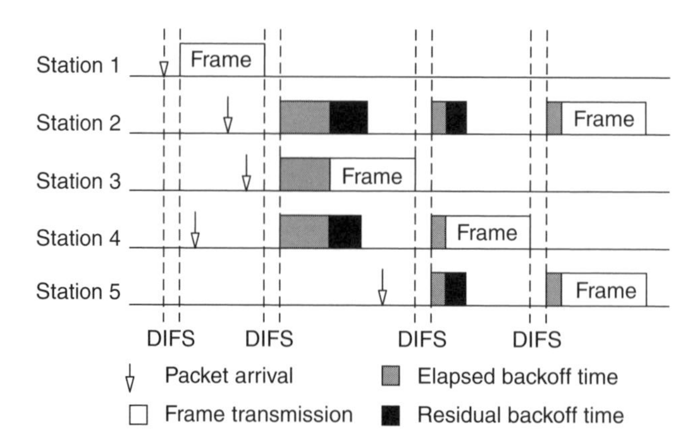

Figure 3.3 Basic access mechanism.

be idle for a time interval greater than a Distributed Interframe Space (DIFS), the station transmits the packet. Otherwise, the transmission is deferred and the backoff process is started. The station computes a random time interval, the backoff interval, uniformly distributed between zero and a maximum called the *Contention Window* (CW). This backoff interval is then used to initiate the backoff timer, which is decremented only when the medium is idle, and it is frozen when another station is transmitting. Every time the medium becomes idle, the station waits for a DIFS and then periodically decrements the backoff timer. The decrementing period is the slot time corresponding to the maximum round trip delay between two stations controlled by the same access point.

When the backoff timer expires, the station can access the medium. If more than one station starts transmission simultaneously, a collision occurs. In a wireless environment, collision detection is not possible. A positive acknowledgement ACK shown in Figure 3.4 is used to notify the sending station that the transmitted frame was successfully received. The transmission of the ACK is initiated at a time interval equal to the Short Interframe Space (SIFS) after the end of reception of the previous frame. The SIFS is shorter than DIFS; thus the receiving station does not need to sense the medium before transmitting the ACK.

If the ACK is not received, the station assumes that the transmitted frame was not successfully received, and it schedules a retransmission and enters the backoff process

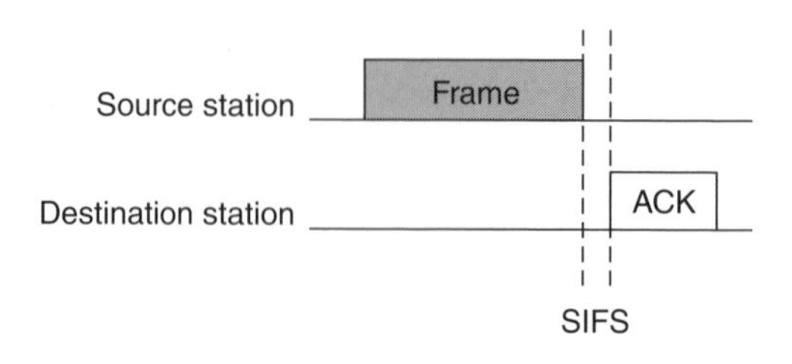

Figure 3.4 Acknowledgement mechanism.

again. After each unsuccessful transmission attempt, the CW is doubled until a predefined maximum (CW_{max}) is reached. This reduces the probability of collisions. After a successful or unsuccessful frame transmission, the station must execute a new backoff process if there are frames queued for transmission.

The hidden station problem occurs when a station successfully receives frames from two different stations that cannot receive signals from each other. This may cause a station to sense the medium being idle even if the other station is transmitting. This results in a collision at the receiving station. The IEEE 802.11 MAC protocol includes an optional mechanism based on the exchange of two short control frames, as shown in Figure 3.5, to solve the hidden station problem. A Request To Send (RTS) frame is sent by a potential transmitter to the receiver. A Clear To Send (CTS) frame is sent by the receiver in response to the received RTS frame. If the CTS frame is not received within a predefined time interval, the RTS frame is retransmitted by executing the backoff algorithm. After a successful exchange of RTS and CTS frames, the data frame is sent by the transmitter after waiting for a SIFS.

A duration field in RTS and CTS frames specifies the time interval necessary to completely transmit the data frame and the related ACK. This information is used by the stations that hear either the transmitter or the receiver to update their Net Allocation Vector (NAV), a timer that is continuously decremented regardless of the status of the medium. The stations that hear either the transmitter or the receiver refrain from transmitting until their NAV expires, and the probability of a collision occurring because of a hidden station is reduced. The RTS/CTS mechanism introduces an overhead that may be significant for short data frames. When RTS/CTS mechanism is enabled, collisions can occur only during the transmission of the RTS frame, which is shorter than the data frame. This reduces the time of collision and wasted bandwidth.

The effectiveness of the RTS/CTS mechanism depends on the length of the data frame to be protected. The RTS/CTS mechanism improves the performance when data frame sizes are larger than the size of the RTS frame, which is the RTS threshold. The RTS/CTS mechanism is enabled for data frame sizes over the threshold and is disabled for data frame sizes under the threshold.

To support time-bounded services the IEEE 802.11 standard defines the PCF to allow a single station in each cell to have a priority access to the medium. This is implemented by using the PCF Interframe Space (PIFS) and a beacon frame that notifies all the other

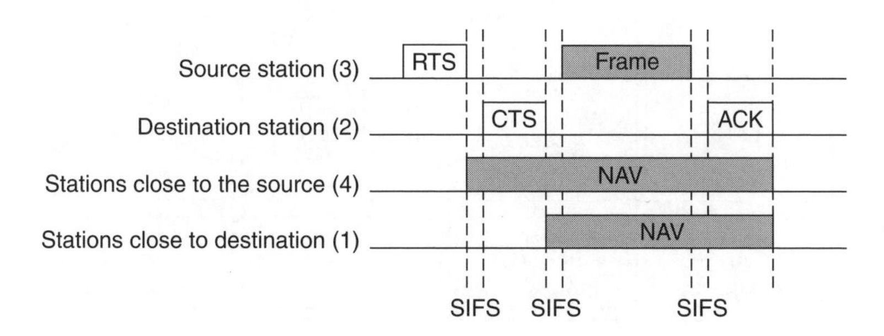

Figure 3.5 Request To Send/Clear To Send (RTS/CTS) mechanism.

stations in the cell not to initiate transmissions for the length of the Contention-Free Period (CFP). When all the stations are silenced, the PCF station allows a given station to have contention-free access by using an optional polling frame sent by the PCF station. The length of the CFP can vary within each CFP repetition interval, depending on the system load.

3.2.5 ETSI HIPERLAN

HIPERLAN standards defined by ETSI are high performance radio LANs. There are four HIPERLAN types illustrated in Figure 3.6 with the operating frequencies and indicative data transfer rates on the radio interface.

In HIPERLAN Type 1, which is also Wireless 8802 LAN, the HIPERLAN Channel Access Mechanism (CAM) is based on channel sensing and a contention resolution scheme called *Elimination Yield – Non-preemptive Priority Multiple Access* (EY-NPMA). The channel status is sensed by each station in the network. If the channel is sensed as being idle for at least 1700 bit periods, the channel is considered free, and the station is allowed to start transmission of the data frame. Each data frame transmission must be acknowledged by an ACK from the destination station.

If the channel is not free when a frame transmission is desired, a channel access with synchronization takes place. Synchronization is performed at the end of the previous transmission interval, and the channel access cycle begins according to the EY-NPMA scheme. The channel access cycle consists of three phases: prioritization, contention, and transmission. Figure 3.7 shows an example of a channel access cycle with synchronization.

Prioritization phase is used to allow only contending stations with the highest priority frames to participate in the next phase. A CAM priority level h is assigned to each frame. Priority levels are numbered from 0 to $(H - 1)$, where 0 is the highest priority level. The prioritization phase consists of at most H prioritization slots, each 256 bit periods long. During priority detection, each station that has a frame with CAM priority level h senses the channel for the first h prioritization slots. In priority assertion, if the channel is idle during this interval, the station transmits a burst in the $(h + 1)$th slot, and it is admitted to the contention phase. Otherwise, it stops contending and waits for the channel access cycle. The contention phase starts immediately after transmission prioritization burst and consists of two further phases – elimination and yield.

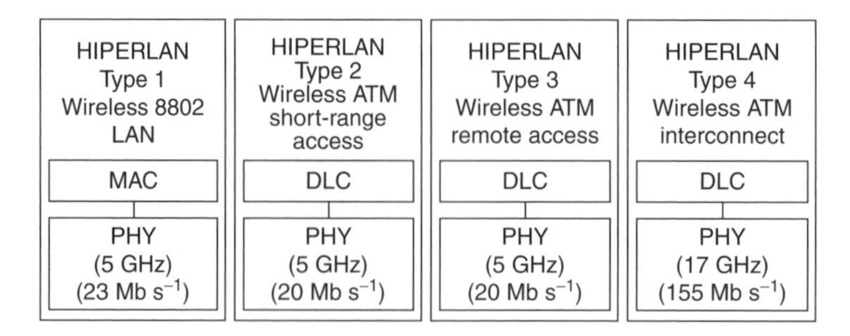

Figure 3.6 HIPERLAN types.

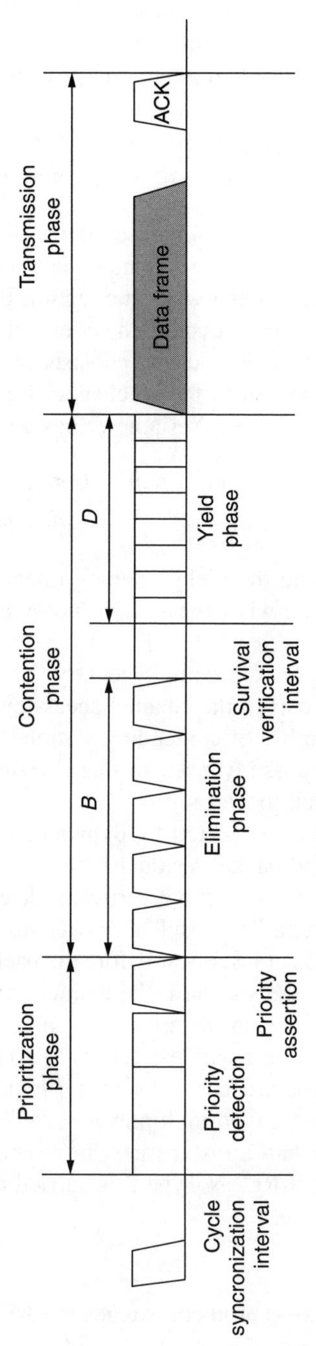

Figure 3.7 Channel access cycle with synchronization.

The elimination phase consists of at most n elimination slots, each 256 bit periods long, followed by a 256–bit period–long elimination survival verification slot. Beginning with the first elimination slot, each station transmits a burst for a number B of elimination slots, according to the following truncated geometric probability distribution function:

$$Pr\{B = b\} = \begin{cases} (1-q)q^b & 0 \leq b < n \\ q^n & b = n \end{cases}$$

When burst transmission ends, each station senses the channel for the duration of the elimination survival verification slot. If the channel is sensed as being idle, the station is admitted to the yield phase. Otherwise, the station drops itself from contention and waits for the next channel access cycle. The yield phase starts after the end of the elimination survival verification interval and consists of at most m yield slots, each 64–bit periods–long. Each station listens to the channel for a number D of yield slots before beginning transmission, if allowed. Variable D has a truncated geometric distribution function:

$$Pr\{D = d\} = \begin{cases} (1-p)p^d & 0 \leq d < m \\ p^m & d = m \end{cases}$$

If the channel is sensed idle during the yield listening interval, the station is allowed to begin the transmission phase. Otherwise, the station looses contention and waits for the next channel access cycle.

The elimination and yield phases are complementary. The elimination phase reduces the number N of stations taking part in the channel access cycle. The yield phase, which performs well with the small number of contending stations, further reduces the number of stations allowed to transmit, possibly even to one. Furthermore, with EY-NPMA at least one station is always allowed to transmit.

Real-time traffic transmission is supported by dynamically varying the CAM priority depending on the user priority and packet residual lifetime. The user priority is assigned to each packet according to the type of traffic it carries; it determines the maximum CAM priority value the packet can reach. The residual packet lifetime is the time interval in which the transmission of the packet must occur before the packet must be discarded. Since multihop routing is supported by the standard, the residual packet lifetime is normalized to the number of hops the packet has to traverse to reach the final destination.

HIPERLAN Type 2 is a short-range wireless access to ATM networks providing local wireless access to ATM infrastructure networks by terminals that interact with access points connected to an ATM switch or multiplexer. WATM access network provides the QoS, including the required data transfer rates the users expect from a wired ATM network. The specification of HIPERLAN Type 2 is carried out by ETSI BRAN.

3.2.6 Dynamic slot assignment

Dynamic Slot Assignment (DSA++) protocol extends the ATM statistical multiplexing to the radio interface of wireless users. The architecture of ATM multiplexer with radio cell is shown in Figure 3.8. The radio cell has a central BS and Wireless Terminals (WTs),

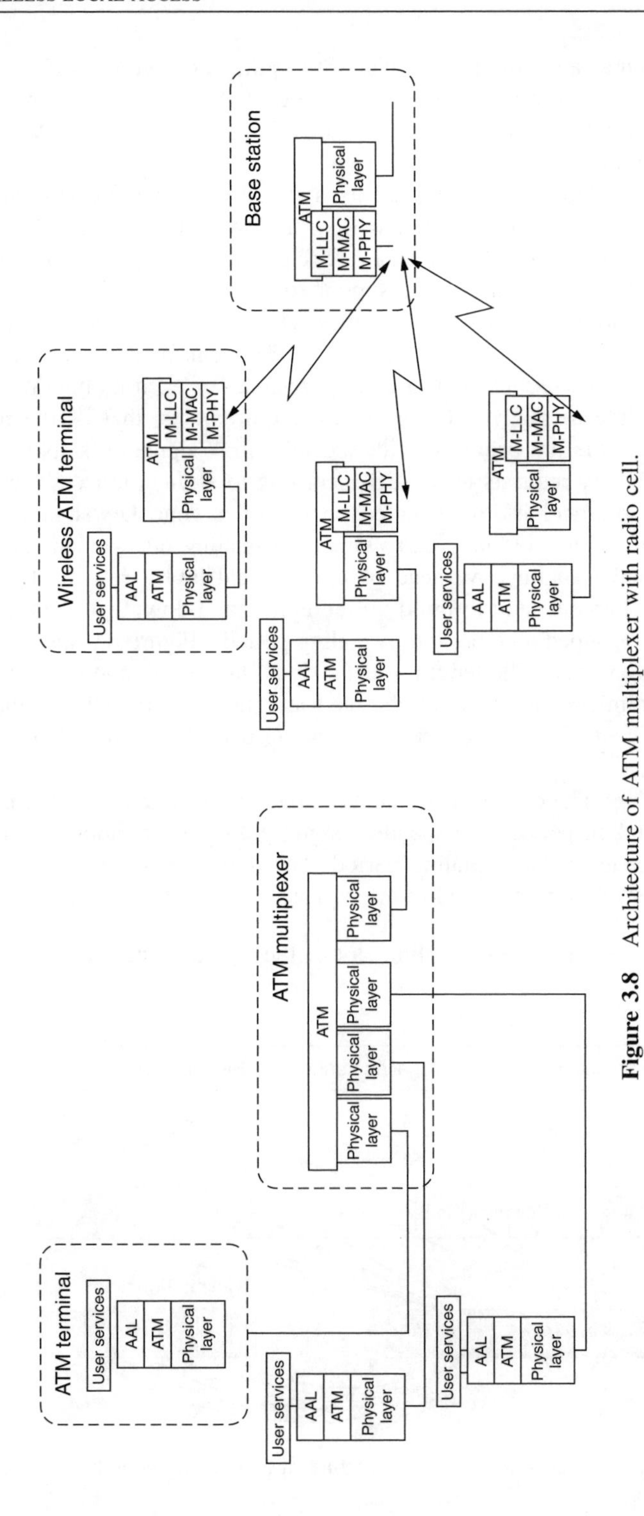

Figure 3.8 Architecture of ATM multiplexer with radio cell.

and can be viewed as a distributed, virtual ATM multiplexer with a radio interface inside. This allows for a centralized master–slave type of MAC protocol, where the BS, as the master of a radio cell, schedules the contention-free transmission of ATM cells on the uplink and downlink.

The virtual ATM multiplexer represents a distributed queuing system with queues inside the WTs for uplink cells and the BS for downlink cells. Similarly, as in fixed ATM networks with a relatively low data rate (e.g., $20\,\text{MB}\,\text{s}^{-1}$), the QoS requirements of real-time oriented services can only be supported if the transmission order of ATM cells is based on the waiting time inside the queues. The BS needs to have current knowledge of the capacity requirements of the mobile WTs. This can be achieved by piggybacking onto uplink ATM cells the instantaneous requirements of each mobile WT. However, it may not be possible to piggyback the newest requirements, that is, the mobile WT is idle. In this case, WTs are provided with special uplink signaling slots so that they can transmit their capacity requests to the BS according to a random access scheme.

The DSA++ protocol is implemented on top of a Time Division Multiple Access (TDMA) channel. Time slots may carry either a signaling burst or one ATM cell along with the additional signaling overhead of the physical layer. A Time Division Duplex (TDD) system is implemented to build up the uplink and downlink channels.

Time slots are grouped together into signaling periods. Figure 3.9 shows a frame structure of a signaling period. The length of each signaling period, and the ratio between the uplink and downlink sections, is variable and assigned dynamically by the BS to cope with the current load of the system. Each signaling period consists of four phases.

Downlink signaling: The downlink signaling burst is transmitted from the BS to the WTs and opens a signaling period of a specific length, giving information about the structure and slot assignments of the signaling period. The downlink signaling informs the WTs about the number of slots in the other three phases and contains at least

- a reservation message for each uplink slot of the signaling period;

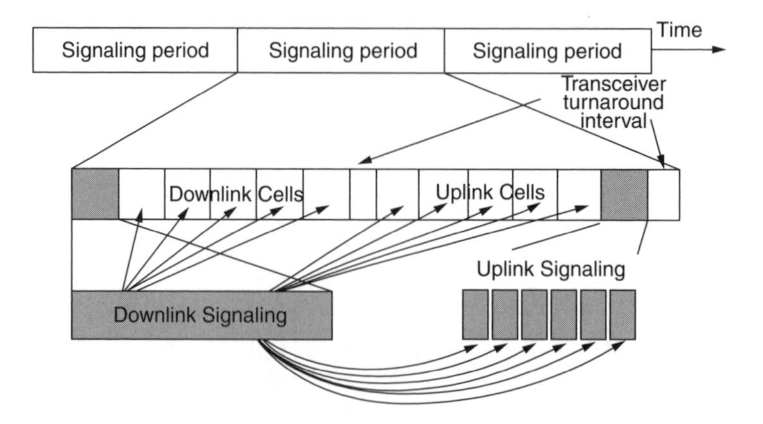

Figure 3.9 Frame structure of a signaling period.

- an announcement message for each downlink slot of the signaling period;
- a control message to implement the collision resolution algorithm of the random access.

Downlink cells: In this phase the downlink cells are transmitted contention-free from the BS to the WTs.

Uplink cells: Since each of these slots is assigned to specific WTs, in this phase uplink cells are transmitted contention-free from the WTs to the BS.

Uplink signaling: During this phase, which is carried out *via* a sequence of short slots, the WTs have the possibility to access the channel to signal their capacity requests to the BS.

Random access is used for transmission of the capacity requests of the WTs. To guarantee the QoS requirements of the connections, fast collision resolution with a deterministic delay is essential. Since all WTs are the possible candidates to transmit *via* random access and are known by the BS, an identifier splitting algorithm can be used, which leads to short and deterministic delays to resolve any collision. The splitting algorithm groups the terminals into sets. All terminals in a set are allowed to transmit in a specific slot. A transmission will only be successful if exactly one terminal in a set transmits. If a collision occurs, the set is divided into subsets according to the order of the splitting algorithm. In the case of an identifier splitting algorithm, the follow-up subset is determined by the identifier of the terminal. An example of a binary identifier splitting algorithm with an identifier space of dimension $n = 4$ is shown in Figure 3.10, where τ_p is the duration of a period able to offer any random access slots.

In DSA++ protocol, at the beginning of each frame the identifier space of size N is divided into a variable number t of consecutive intervals and a random access slot

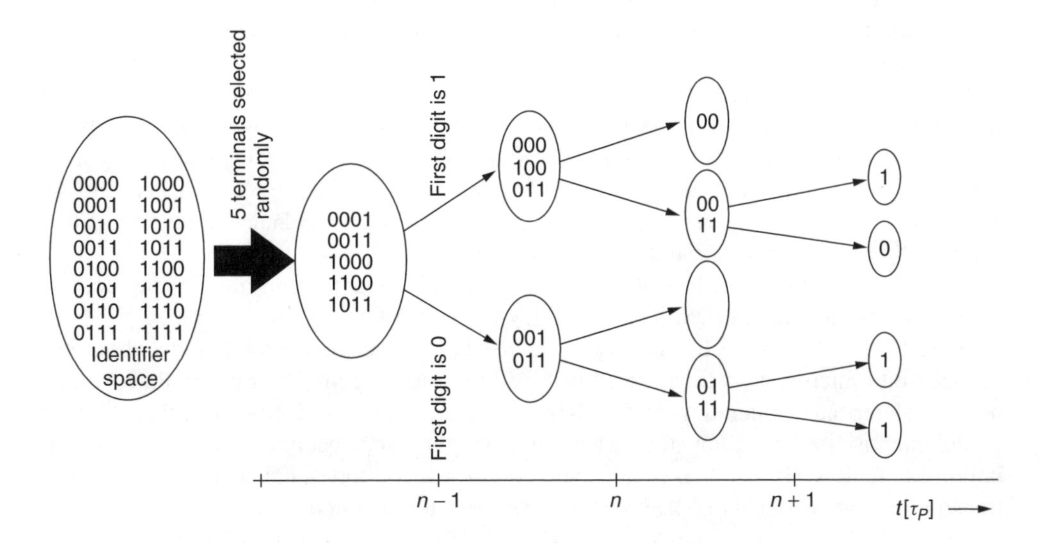

Figure 3.10 An example of a binary identifier splitting algorithm.

is assigned to each interval. The lth interval starts with terminal i_l and ends with terminal $(i_{\{l+1\}} - 1)$, with $i_1 = 0$, and $i_t = (N - 1)$. The downlink signaling burst signals the interval division to the WTs by transmitting the start identifier i_l of each interval. The maximum time required to resolve the collision is limited because of the limited and known number of WTs served by the BS. Petras and Kramling show that the solution time of a collision can be reduced by using an estimate of the transmission probability of each terminal to determine the size of the subsets and the splitting order.

The coding of the capacity requests and the scheduling algorithm depend on the ATM-service class. An earliest due date strategy is used for Constant Bit Rate (CBR) and real-time Variable Bit Rate (rt-VBR) service classes. For Available Bit Rate (ABR) and Unspecified Bit Rate (UBR) service classes, Fair Weighted Queuing and First Come First Served (FCFS) strategies are used.

3.3 SUMMARY

In IPv6, a special address range is reserved for multicast addresses for each scope, and a multicast is only received by the hosts in this scope, which are configured to listen to this specific multicast address. To address all hosts in a certain scope with a multicast, the multicast must be made to the predefined all-nodes address, to which all hosts must listen. When existing software using IPv4 is migrated to IPv6, the IPv4 broadcasts are changed to multicasts to the all-nodes address, as this is the simplest way to maintain the complete functionality of the software.

In a workgroup address configuration, the host sends a DHCP Request with a Workgroup Address Extension to the DHCP Server. The DHCP Server replies with a Workgroup Address Extension containing all workgroup addresses assigned to this host. After receiving the workgroup addresses, the host sends ICMPv6 Group Membership Report to each of its workgroup addresses to inform the multicast routers about its new membership in these multicast groups.

OFDM modulation combined with DPA with wideband 5-MHz channels for high-speed packet data wireless access in macrocellular and microcellular environments supports bit rates ranging from 2 to 10 $\mathrm{Mb\,s^{-1}}$. OFDM can largely eliminate the effects of intersymbol interference for high-speed transmission rates in very dispersive environments. OFDM supports interference suppression and space–time coding to enhance efficiency. DPA supports spectrum efficiency and high-rate data access.

Several systems support broadband wireless communications and mobile user access. These are MMDS and LMDS, also called LMCS or MVDS.

Broadband wireless access is based on the TLN concept in which subscribers are grouped into microcells, which are embedded into a macrocell. The microcells coverage uses local repeaters operating at 5.8 GHz fed by a BS through 40-GHz links. OFDM modulation is used to allow the reception with plug-free receivers located inside the buildings. A 40-GHz band fixed receiver provides a rooftop antenna in LOS with the transmitting antenna. This LMDS system provides an integrated wireless return channel.

IEEE 802.11 uses data rates up to 2 $\mathrm{Mb\,s^{-1}}$ and defines two network topologies. The infrastructure-based topology allows MTs to communicate with the backbone network through an access point. In *ad hoc* topology, MTs communicate with each other without connectivity to the wired backbone network. HIPERLAN uses data rate $23.5\,\mathrm{Mb\,s^{-1}}$ and the *ad hoc* topology.

DSA++ protocol extends the ATM statistical multiplexing to the radio interface of wireless users. The architecture of ATM multiplexer with radio cell has a central BS and WTs, and can be viewed as a distributed, virtual ATM multiplexer with a radio interface inside. This allows for a centralized master-slave type of MAC protocol, in which the BS, as the master of a radio cell, schedules the contention-free transmission of ATM cells on the uplink and downlink.

PROBLEMS TO CHAPTER 3

Wireless local area networks

Learning objectives

After completing this chapter, you are able to

- demonstrate an understanding of virtual LANs;
- explain the role of workgroups;
- explain multicasting in virtual LANs;
- explain workgroup address configuration;
- demonstrate an understanding of OFDM;
- explain what WCDMA is;
- explain DPA;
- demonstrate an understanding of LMDS;
- explain what MMDS is;
- explain what HFR, RTTB, and RTTC are;
- demonstrate an understanding of different MAC protocols for wideband wireless local access;
- explain what IEEE 802.11 and HIPERLAN standards are;
- explain what Dynamic Slot Assignment (DSA++) protocol is;

Practice problems

3.1: What are the workgroups?
3.2: How is multicasting done in IPv6?
3.3: How is administration of workgroups designed?
3.4: What peak bit rates are supported by OFDM?
3.5: What is the role of WCDMA?
3.6: What is the function of DPA?
3.7: What is the role of BRAN?
3.8: What can the MMDS systems be used for?
3.9: What is the coverage for LMDS systems?
3.10: How does the user access the network?

3.11: What are the services provided by the IEEE 802.11 MAC?

3.12: How does the CAM work in HIPERLAN Type 1?

3.13: How does the DSA++ protocol extend the ATM statistical multiplexing?

Practice problem solutions

3.1: The workgroups are groups of hosts sharing the same servers and other resources over the network. The hosts of a workgroup are attached to the same LAN segment, and broadcasting can be used for server detection, name resolution, and name reservation.

3.2: In IPv6, a special address range is reserved for multicast addresses for each scope, and a multicast is only received by the hosts in this scope, which are configured to listen to this specific multicast address. To address all hosts in a certain scope with a multicast, the multicast must be made to the predefined all-nodes address, to which all hosts must listen. When existing software using IPv4 is migrated to IPv6, the IPv4 broadcasts are changed to multicasts to the all-nodes address, as this is the simplest way to maintain the complete functionality of the software.

IPv6 multicasting can be used to form the broadcast scope of a workgroup. The workgroup is the multicast group, whose hosts listen to the same multicast address, the workgroup address. A host can listen to several multicast addresses at the same time and can be a member of several workgroups.

Multicasting exists optionally for IPv4 and is limited by a maximum of hops. The multicast in IPv6 is limited by its scope, which is the address range.

3.3: The administration of the workgroups is designed by storing the information about hosts and their workgroups in a central database in a DHCP server. The information is distributed by using the DHCPv6.

3.4: OFDM modulation combined with DPA with wideband 5-MHz channels for high-speed packet data wireless access in macrocellular and microcellular environments, supports peak bit rates ranging from 2 to 10 $\mathrm{Mb\,s^{-1}}$.

3.5: WCDMA uses 5-MHz channels and supports circuit and packet data access at 384 $\mathrm{kb\,s^{-1}}$ nominal data rates for macrocellular wireless access. WCDMA provides simultaneous voice and data services.

3.6: DPA is based on properties of an OFDM physical layer. DPA reassigns transmission resources on a packet-by-packet basis using high-speed receiver measurements.

3.7: BRAN provides a high-speed digital connection to the user.

3.8: The MMDS systems work at frequencies lower than 5 GHz in large coverage areas with cell radius of up to 40 km. MMDS systems can be used for transmission of video and broadcast services in rural areas. Because of a large cell size, MMDS systems do not perform well for bidirectional communication that integrates a return channel.

3.9: The LMDS systems work with higher frequencies where larger frequency spectrum is available than that in the MMDS systems. The coverage for LMDS systems involves smaller cells of up to 5-km radius, and requires repeaters to be placed in a LOS configuration. This local coverage with a large available bandwidth makes LMDS systems suitable for interactive multimedia services distribution.

3.10: The user accesses the network through HFR, RTTB, and RTTC. In HFR an RFU carries out signal down conversion from RF frequency to the intermediate frequency. The signal feeds the RT of each user through a bus link. In RTTB architecture the signal feeds the user NT through point-to-point cable links. In RTTC the RFU is placed in a common outdoor unit and is shared among several buildings.

3.11: The IEEE 802.11 MAC (Media Access Control) protocol provides asynchronous and synchronous (contention-free) services, which are provided on top of physical layers and for different data rates. The asynchronous service is mandatory, and the synchronous service is optional.

3.12: In HIPERLAN Type 1, which is also a Wireless 8802 LAN, the HIPERLAN CAM is based on channel sensing and a contention resolution scheme called *EY-NPMA*. The channel status is sensed by each station in the network. If the channel is sensed as being idle for at least 1700 bit periods, the channel is considered free, and the station is allowed to start transmission of the data frame. Each data frame transmission must be acknowledged by an ACK from the destination station.

If the channel is not free when a frame transmission is desired, a channel access with synchronization takes place. Synchronization is performed at the end of the previous transmission interval, and the channel access cycle begins according to the EY-NPMA scheme. The channel access cycle consists of three phases: prioritization, contention, and transmission.

3.13: DSA++ protocol extends the ATM statistical multiplexing to the radio interface of wireless users. The architecture of ATM multiplexer with radio cell has a central BS and WTs and can be viewed as a distributed, virtual ATM multiplexer with a radio interface inside. This allows for a centralized master-slave type of MAC protocol, in which the BS, as the master of a radio cell, schedules the contention-free transmission of ATM cells on the uplink and downlink.

4

Wireless protocols

A MAC protocol for a wireless LAN provides two types of data-transfer Service Access Points (SAP): network and native. The network SAP offers an access to a legacy network protocol (e.g., IP). The native SAP provides an extended service interface that may be used by custom network protocols or user applications capable of fully exploiting the protocol specific Quality of Service (QoS) parameters within the cell service area.

Broadband Radio Access Integrated Network (BRAIN) is used for millimeter wave band multimedia communications. In BRAIN, all Access Points (APs) need to have only an optical/electrical (OE) converter because BRAIN incorporates radio on fiber technologies, which allow for transmitting radio signals through optical fiber cables.

The Hybrid and Adaptive MAC (HAMAC) protocol integrates fixed assignment Time Division Multiple Access (TDMA) protocols, reservation-based protocols, and contention-based protocols into a wireless network, simultaneously and efficiently supporting various classes of traffic such as Constant Bit Rate (CBR), Variable Bit Rate (VBR), and Available Bit Rate (ABR) traffic. The HAMAC protocol uses a preservation slot technique to minimize the packet contention overhead in Packet Reservation Multiple Access (PRMA) protocols, while retaining most isochronous service features of TDMA protocols to serve voice and CBR traffic streams.

Adaptive Request Channel Multiple Access (ARCMA) is a Demand Assignment Multiple Access (DAMA) protocol with dynamic bandwidth allocation. This scheme is designed to function in a cell-based wireless network with many Mobile Stations (MSs) communicating with the Base Station (BS) of their particular cell. Transmissions are done on a slot-by-slot basis without any frames. Each slot is divided into a Transmission Access (TA) slot and a Request Access (RA) minislot. The RA channel in ARCMA is capable of carrying additional information for different classes of Asynchronous Transfer Mode (ATM) service (e.g., CBR, VBR, etc.). This additional information is used by the BS to provide better QoS support for different classes of traffic. Transmission from CBR traffic may reserve an incremental series of slots in the duration of their transmission. No further request is needed until the CBR transmission finishes.

4.1 WIRELESS PROTOCOL REQUIREMENTS

The general requirements for wireless protocols supporting wireless LANs are as follows:

- The low cost is achieved by simple implementation and the use of standard multipurpose modules and components. Modularity and reconfigurability in all stages of system design are the key elements to meet these requirements.
- The QoS requirements for the data-transfer service of the MAC protocol include support for user-defined traffic types and connection parameters. The protocol must support real-time data-transfer services.
- The wireless LAN can be used both as an extension and as an alternative to a wired LAN. Therefore, for interoperability requirements, the changing topology of a wireless network, inadequate security and reliability of the medium, and protocol-specific management functionality must be hidden from the network user, that is, from legacy Transmission Control Protocol/Internet Protocol (TCP/IP) applications.
- Wireless medium does not provide the same level of confidentiality and user identification as a wired system. A wireless coverage area cannot be reliably defined or restricted. Actions at the MAC layer have to be taken to provide a secure data-transfer service.
- An unlicensed and globally available frequency band must be selected for the system.
- The architecture of the MAC protocol should follow a master–slave hierarchy as the centralized control and management enables an easy and efficient support of QoS parameters and an access point for outside network resources.
- To guarantee the low cost, efficient resource management and guaranteed QoS, the number of simultaneous users in a single wireless LAN cell can be restricted according to the target environment.
- The requirement for low power consumption follows from the usage of battery powered portable network equipment, for example, laptops. A wireless network adapter should not significantly shorten the operating time of a portable terminal. Therefore, the MAC protocol should be capable of turning off the transceiver during idle periods without missing any relevant transmission.

4.2 MAC PROTOCOL

Hannikainen *et al.* present a MAC protocol for a wireless LAN that provides two types of data-transfer SAP: network and native. The network SAP offers an access to a legacy network protocol (e.g., IP). The native SAP provides an extended service interface that may be used by custom network protocols or user applications capable of fully exploiting the protocol specific QoS parameters within the cell service area.

The data processing block converts the user data into a more suitable form for the wireless medium. Encryption is performed for confidentiality while fragmentation and Forward Error Correction (FEC) coding functions are added for better protection of the data against transmission errors. The frame queuing and Automatic Repeat Request (ARQ) retransmissions are controlled according to assigned QoS.

The control and management functionality consists of state machines that adapt to the inputs from the management interface and data processing modules while producing an output according to the current state of the system. The operational parameters are stored in the Management Information Base (MIB), which can be accessed and modified through the station management-user interface.

Both the Portable Station (PS) and BS functionality are assembled using the same functional modules. The BS functionality is achieved by adding the base-specific modules (data processing, control, and management) on top of the PS functionality. A set of BS functions can be included into a PS capable of executing them (for instance, a laptop). Thus, an *ad hoc* networking is enabled if no permanent BS service is available.

Hannikainen *et al.* present a connection-oriented wireless MAC protocol that uses a reservation-based TDMA scheme with the shared medium. The medium access cycle is divided into time slots that form CBR channels. Four types of channels are distinguished by their purpose, direction, and bandwidth. These are data, contention, control, and beacon channels. The data channel includes also acknowledgements.

The data channel can be reserved by a PS for uplink transmission of user data. The data is forwarded by the BS; however, a direct data transfer between two PS under a BS control is possible. The data channels remain reserved during data transmission and are released by the PS request, or by the BS in the case of an idle reserved channel.

Another uplink control channel is formed by the Acknowledgement (ACK) messages that follow each unicast transmission destined to a single station. The protocol uses a store and wait flow-control scheme to enable short retransmission delays and fast adaptation to the varying quality of radio link. The acknowledgements carry information about successful or unsuccessful reception, and control information for bandwidth requirements, which consists of the amount and priority of data queued in a PS for transmission. This information is used by the channel-scheduling function of a BS to determine the uplink or downlink direction of the reserved data channels and the possible requirements for an extra bandwidth for each PS.

The PS transmits uplink control messages, such as channel reservation and association requests, in a contention-based channel at the end of the access cycle. The contention channel is constructed by a series of short contention slots that are monitored for a signal carrier or energy for Carrier Sense Multiple Access (CSMA)–based transmissions. The amount of idle contention slots to be detected before transmitting enables various control message priorities.

The BS transmits downlink control messages in the control and beacon channels. The control channel is a data channel that can be reserved for control and management information transfer. Otherwise, the channel is used as a VBR data channel. The beacon channel is used by the BS only for beacon messages. A beacon broadcasts the current channel reservation state for the following access cycle. Beacons also carry cell-specific information, such as a cell identification, structure of the access cycle, and indications for a required confidentiality with the association and data transfer. A beacon frame indicates the beginning of the access cycle, thus providing a TDMA cycle synchronization for PS. The beacon carries indications for buffered data to those PS that use power-save functionality. These stations power on their receivers only periodically to receive the possible announcement with the beacon.

4.3 BROADBAND RADIO ACCESS INTEGRATED NETWORK

Inoue *et al.* present the BRAIN for millimeter wave band multimedia communications. In BRAIN, all APs need to have only an OE converter because BRAIN incorporates radio-on-fiber technologies that allow for transmitting radio signals through optical fiber cables.

Reservation Based Slotted Idle Signal Multiple Access (RS-ISMA) is a wireless access protocol designed for wireless multimedia communications and implemented in the BRAIN indoor-LAN prototype. In addition, a compact Radio Frequency (RF) module composed of flat antennas and a Monolithic Microwave Integrated Circuit (MMIC) was employed for each remote station and AP. The use of large capacity Field Programmable Gate Arrays (FPGA) decreased the number of signal processing boards. System parameters such as the packet format were optimized for Internet Protocol (IP) datagram transport to support all applications based on IP. The function of Negative Acknowledgement (NACK) sensing was added to RS-ISMA to ensure an efficient and smooth wireless multicast in a multiple access environment.

BRAIN covers service areas with multiple Basic Service Area (BSA), which includes an AP and a number of fixed and/or quasi-fixed stations (ST). ST usually employs a directional antenna in millimeter wave band communications and communicates *via* AP. Traffic generated from or arriving at the BSA passes through the AP, and thus the indoor system is a centralized control system.

RS-ISMA is a wireless MAC protocol that is an integration of reservation-based ISMA and slotted ISMA, and it is basically a combination of random access protocol and polling protocol. During the reservation step, an ST transmits a short frame to make a reservation under a random access scheme. In the information transmission step, either an isochronous or an asynchronous polling scheme is used for information transmission depending on the QoS requirements.

RS-ISMA was modified to carry IP datagram most efficiently and to support wireless multicast. The MAC frame format has a fixed length to increase the signal processing speed, resulting in increased radio transmission speed. The payload of modified RS-ISMA is 64 octets and the header is 4 octets.

A Stop and Wait (SW) ARQ with a limited number of retransmissions is used for both stream traffic and data traffic in the modified RS-ISMA because combining the TCP error-recovery mechanism with SW ARQ allows for low-frame error rate necessary for reliable transmission of data traffic.

In multimedia wireless LAN, the retransmission scheme for downlink frame transmission should enable broadcast and multicast of multimedia traffic to multiple users without errors. The AP, after sending a data frame, transmits a polling signal whose control signal field indicates Acknowledgement Request (ACKR). In response to the ACKR frame station, ST1 does not send a frame because it has received the data frame successfully. Station ST2 sends an ACK frame that informs the AP that the data frame has not been received successfully. The AP, which senses any carrier from STs during one time slot following the ACKR, detects a carrier from station ST2. The AP generally does not know which ST has transmitted a signal since there may be more than two STs that are sending

ACK frames because they have not received the downlink data frame without errors. After detecting any carrier, the AP retransmits the data frame, which will be received by both stations ST1 and ST2 but will be ignored by ST1.

4.4 HYBRID AND ADAPTIVE MAC PROTOCOL

Wang and Hamdi propose a MAC protocol HAMAC, which, integrates fixed assignment TDMA protocols, reservation-based protocols, and contention-based protocols into a wireless network, simultaneously and efficiently supporting various classes of traffic such as CBR, VBR, and ABR traffic. The HAMAC protocol uses a preservation slot technique to minimize the packet contention overhead in PRMA protocols, while retaining most isochronous service features of TDMA protocols to serve voice and CBR traffic streams.

The HAMAC protocol uses a super frame that is divided into two frames, the downlink frame and the uplink frame. The length of the frames can vary depending on the bandwidth demand. The downlink frame is used by the BS to broadcast the frame configuration information, the connection setup, the allocation information, the request information, and the data to all mobile devices. The information and the data can be broadcast using a single burst because only the BS controls the downlink. Mobile devices can filter out irrelevant information upon receiving them. The first segment of the downlink frame is used for control signaling needed for the frame configuration to be known by all mobile devices before starting the reception and the transmission.

In the HAMAC protocol, the uplink frame consists of three segments. The first segment is used by the mobile devices to upload the CBR data using a TDMA round-robin scheme. There are two types of slots in this segment: the preservation slot and the normal slot. The preservation slot is used to preserve the position for a CBR connection when it is in a silent state. The length of the preservation slot should be as short as possible. During the transmission of the preservation slot, all mobile devices in the same cell should have enough time to recognize the existence of preservation slot or the existence of silent CBR connection. The preservation slot is not useful for the BS, and it is discarded by the BS and does not appear in the downlink frame.

When the preservation slot of a CBR connection is present, the remaining bandwidth of the connection is free. When the CBR connection becomes active again, the preservation slot is replaced by the normal slots and the allocated bandwidth for the connection cannot be used by the other connections and mobile devices. The HAMAC protocol avoids the reservation operation before the transmission of an active talk spurt, and the BS is not aware of the state transition of the CBR connection. As a result, there is no need to make the presence or absence of the preservation slot known to mobile devices using a downlink frame. The preservation slot can appear or disappear without any notification.

The HAMAC protocol uses the continuous bit to compress the header information of consecutive slots when they belong to the same traffic source. In the continuous bit technique, the position of the slots allocated to the connections can float in the uplink frame, rather than having the slots allocated to a connection being assigned to a fixed location. In HAMAC protocol, the location of the slots allocated to the connection, defined as an access point, is assigned as the function of the number of continuous bits rather than the absolute position relative to the beginning of the super frame. As a result, the whole frame is used efficiently without any unusable fragments left. The location should be adjusted once a CBR connection is dropped, or a new CBR connection is established.

The second segment of the frame in HAMAC protocol is used to carry bursty data packets, which have to be reserved and allocated by the BS scheduler. Bursty data traffic occur in large volumes; thus this segment-frame contains only the normal slots.

The third segment of the frame contains the contention slots only. The contention slots are small minislots to reduce the overhead caused by collisions. These slots are contended for under the control of a permission probability with respect to different types of packets. Reservation packets and control packets are more important since they may affect the performance of the second segment access or they may be network-management packets that need to be served as fast as possible. Hence, they are assigned a higher permission probability. The ABR data packets should not significantly affect the system performance, and they are given relatively low permission probability to contend for the minislots. To ensure that there is always a chance for reservation packets and control packets to transmit, the minimum length is set for the third segment frame.

4.5 ADAPTIVE REQUEST CHANNEL MULTIPLE ACCESS PROTOCOL

ARCMA is a multiple-access protocol based on demand assignments. This scheme is based on the Distributed Queuing Request Update Multiple Access (DQRUMA) protocol and incorporates the periodic traffic handling of PRMA. In addition, ARCMA reduces collisions in the RA channel by using an efficient adaptive request strategy.

ARCMA is a DAMA protocol with dynamic bandwidth allocation. This scheme is designed to function in a cell-based wireless network with many MSs communicating with the BS of their particular cell. Transmissions are done on a slot-by-slot basis without any frames. As with DQRUMA, each slot is divided into a TA slot and an RA-minislot. However, the RA channel in ARCMA is capable of carrying additional information for different classes of ATM service (e.g., CBR, VBR, etc.). This additional information is used by the BS to provide better QoS support for different classes of traffic. As in PRMA, transmission from CBR traffic may reserve an incremental series of slots in the duration of their transmission. No further request is needed until the CBR transmission finishes.

The BS maintains a Request Table to keep track of all successful requests and assigns permission to mobiles for transmission at different time slots. In ARCMA protocol, the BS inspects the service class of a request and gives transmission priority to delay-sensitive data (e.g., CBR). As in the DQRUMA protocol, a piggyback (PGBK) bit is used in the uplink channel to reduce contention in the RA channel. This is especially beneficial for bursty traffic.

ARCMA implements a dynamic RA channel similar to that of DQRUMA in which an entire uplink channel can be converted into multiple RA channels. This conversion is done when the Request Table is empty, which in most cases indicates heavy collisions in the request channel. ARCMA uses an algorithm that takes advantage of the random access scheme in the RA channel. We use the slotted ALOHA with Binary Exponential Backoff (BEB) as the random access protocol for ARCMA.

ARCMA improves the spectral efficiency by reducing collisions in the RA channel while improving support for the various classes of ATM services.

ARCMA protocol is composed of a phase similar to DQRUMA's request/acknowledgement phase and a permission/transmission phase. Both these phases are associated with data transmission from the MS to BS. Data transmission from the BS to MS is a straightforward operation in which the BS merely broadcasts the information (data packets) to the entire cell. The destination MS listens to the broadcast channel and retrieves the appropriate data packets (based on the destination Access ID). If the transmission destination does not reside in the same cell, the BS will forward the packet to an ATM switch to be routed to the proper destination.

4.6 REQUEST/ACKNOWLEDGEMENT PHASE

The request is made in the RA channel (RA minislot). The request data packet contains the mobile's b-bit Access ID assigned during setup. In ARCMA protocol, in addition to the Access ID, the request packet also includes the type of service being requested. The protocol provides additional support for periodic traffic (i.e., CBR). Since traffic can be either CBR or non-CBR, only a single bit is required to identify the service type as shown in Figure 4.1. This bit is transmitted together with the request packet in the RA

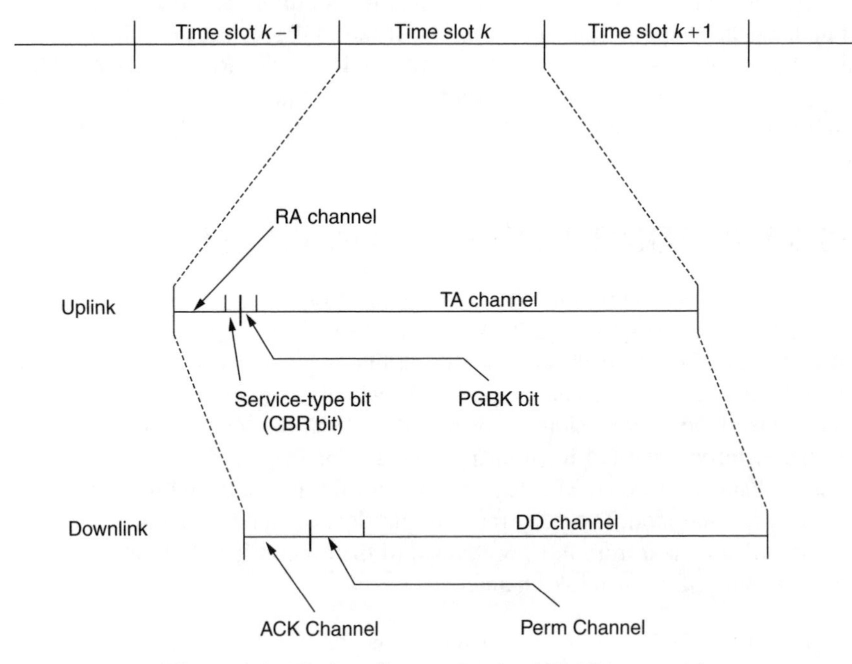

Figure 4.1 Timing diagram for the ARMCA protocol.

channel. DQRUMA provides no distinction between requests of different service types. The distinction provided in ARCMA is used by the BS to assign priority to CBR traffic.

Like most DAMA, the request channel uses a random access protocol for transmission. ARCMA uses the Slotted ALOHA with BEB algorithm. The BEB scheme is a stabilization strategy for protocols with limited feedback such as Slotted ALOHA. When a packet initially arrives at the buffer, a request is transmitted immediately in the next time slot. If there is a collision, the probability of retransmission, q_r is set to a half. If a second collision occurs, q_r is set to a quarter. After i unsuccessful transmissions, the probability of retransmission is given by $q_r = 2^{-i}$. That is, after i collisions, the probability of retransmission is uniformly distributed over the next 2^i time slots. However, we freeze the q_r at 2^{-10} for any retransmission after 10 collisions to prevent the possibility of excessive retransmission delay due to the reduction of q.

After every random transmission, the MS needs to know if the request was successful. Since the MS does not detect collision by sensing the channel, the BS has to send a response to the MS indicating a successful request. When the BS receives a request from the RA channel, it immediately sends (broadcasts) an ACK to the MS. The BS then inserts the new request in a Request Table to indicate that the MS has packet(s) to transmit. The Request Table contains all the unprocessed requests received by the BS. This table is used for scheduling TA. After the MS receives its acknowledgement (by listening to the ACK channel), it listens to the downlink Perm Channel for transmission permission. MS that do not receive acknowledgment for their requests will retransmit their requests according to the slotted ALOHA scheme.

As in DQRUMA, we make use of the PGBK request bit to provide a contention-free request for an MS that has more than one packet in its buffer. The BS checks the PGBK bit and updates the Request Table accordingly. If the PGBK is set to 1, the BS generates a request, for the corresponding MS, to be inserted into the Request Table. That is, a packet arriving at a nonempty buffer does not need a request for TA. There is no ACK associated with the piggyback request. The MS merely listens to the Perm Channel for permission to transmit the next packet in the buffer.

4.7 PERMISSION/TRANSMISSION PHASE

General traffic: The BS is responsible for allocating bandwidth (time slots) to the MS by using a packet transmission policy. In ARCMA protocol, we use a simple, First In First Out (FIFO) policy. The MS that makes a request first is given permission to transmit first. The Request Table is implemented as a queue in which the request goes to the tail of the queue and transmission permission is given to the MS at the head of the queue. The MS that has successfully requested for transmission (RA or PGBK) listens for its Access ID in the Perm Channel. Once an MS hears its Access ID, it is allowed to transmit its data in the following time slot. The MS transmits its data in the TA channel collision-free. The BS forwards the data from the TA channel to the appropriate destination through the Data Downstream (DD) downlink channel.

CBR traffic: The ARCMA protocol provides special handling for CBR traffic. This connection-oriented delay-sensitive traffic is given priority in the request phase. When a CBR request arrives at the BS, it is inserted in special CBR Request Queue. The CBR Request Queue also uses the FIFO policy but requests in this queue have precedence over those in the Request Table. All requests in the CBR Request Queue must be processed prior to those in the Request Table. Since transmission priority is always given to CBR traffic, we must limit the number of MSs with CBR traffic in a cell. Otherwise, general traffic may not be given a chance to transmit. This controlling can be done during call setup when traffic requirements are negotiated.

The transmission rate of CBR traffic is given in the form of arrival rates. This arrival rate depends on the rate of the CBR traffic and the transmission rate of the channel. We assume that CBR generates data packets at a constant rate, hence generating a constant arrival rate throughout the connection. CBR packets arrive at the mobile's buffer every Nth time slot. Since the BS is aware of this, it automatically assigns a time slot, by generating a request in the CBR Request Queue, for that particular MS. No request is required by the MS for the duration of the CBR traffic. The MS only has to listen for its Perm bits before transmitting its data. Each CBR reservation needs to be terminated at the end of its CBR transmission. This is performed using the PGBK bit. CBR transmission does not involve the PGBK bit since consequent time slot allocation is based on the arrival rates. Therefore, the PGBK bit is used by CBR traffic to indicate the end of a CBR transmission. After sending the Perm bits, the BS waits for the CBR packet in the next time slot and checks its PGBK bit. A PGBK bit with a zero value indicates the end of a CBR transmission and the BS will stop assigning periodic time slots for this particular MS. By using the PGBK, no additional data-overhead is needed for the termination procedure.

Figure 4.2 illustrates the general flow of ARCMA protocol at every MS. In ARCMA, only the first CBR data packet has to request an access. The subsequent CBR data packets merely have to listen to the Perm channel for transmission permission. Requests are automatically generated by the BS.

We reduce the collisions in the RA channel by adapting to the traffic environment. We exploit idle TA slots by converting each slot into multiple RA slots. Idle time slots occur when there are no entries in the request queues (general and CBR). In our adaptive scheme, when the BS detects that the request queues are empty, it converts the next uplink (otherwise idle) TA slot into R number of RA minislots as shown in Figure 4.3. The BS does this by sending a multipleRA message in the Perm Channel to all the MSs. In the next time slot (multipleRA mode), MS can randomly select one of the R channels for request transmission. This selection can also be statically assigned by the BS during call setup. To acknowledge these multiple requests, the downlink channel is similarly converted into R number of ACK slots.

While a similar implementation is proposed in DQRUMA, our design reduces idle time slots by considering the probability of retransmission q_r. When the TA channel is first converted into multiple RA minislots, all new and previously unsuccessful requests are transmitted with the probability of 1. That is, all requests are sent out immediately regardless of their q_r. If no requests are successfully transmitted (i.e., request queues remain empty), the uplink channel remains in the multipleRA mode. However, in this case, the MSs retransmit their requests according to their old q_r (based on the BEB algorithm). Conversely, if successful requests were made during the multiple RA slots, the BS reverts

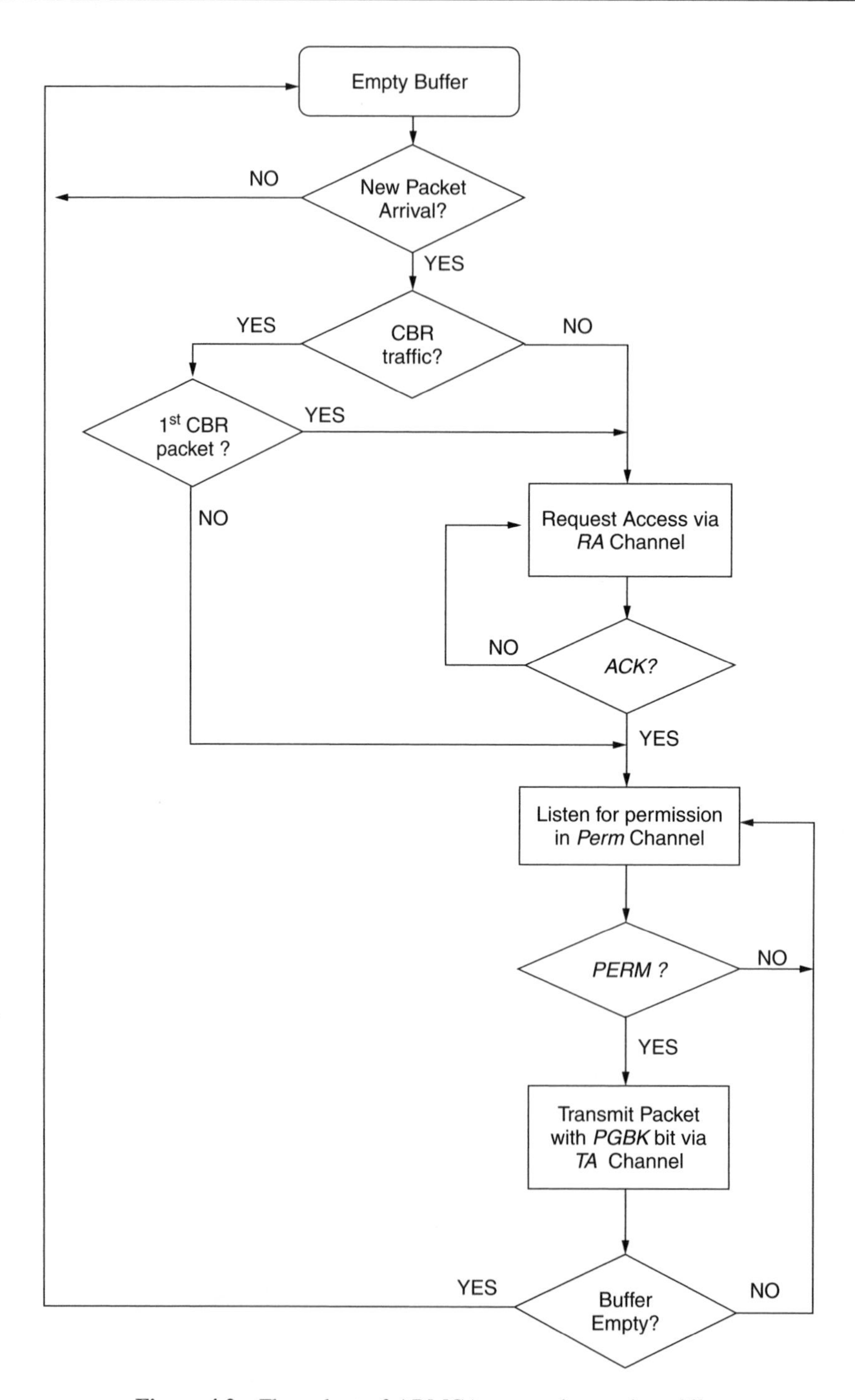

Figure 4.2 Flow chart of ARMCA protocol at each mobile.

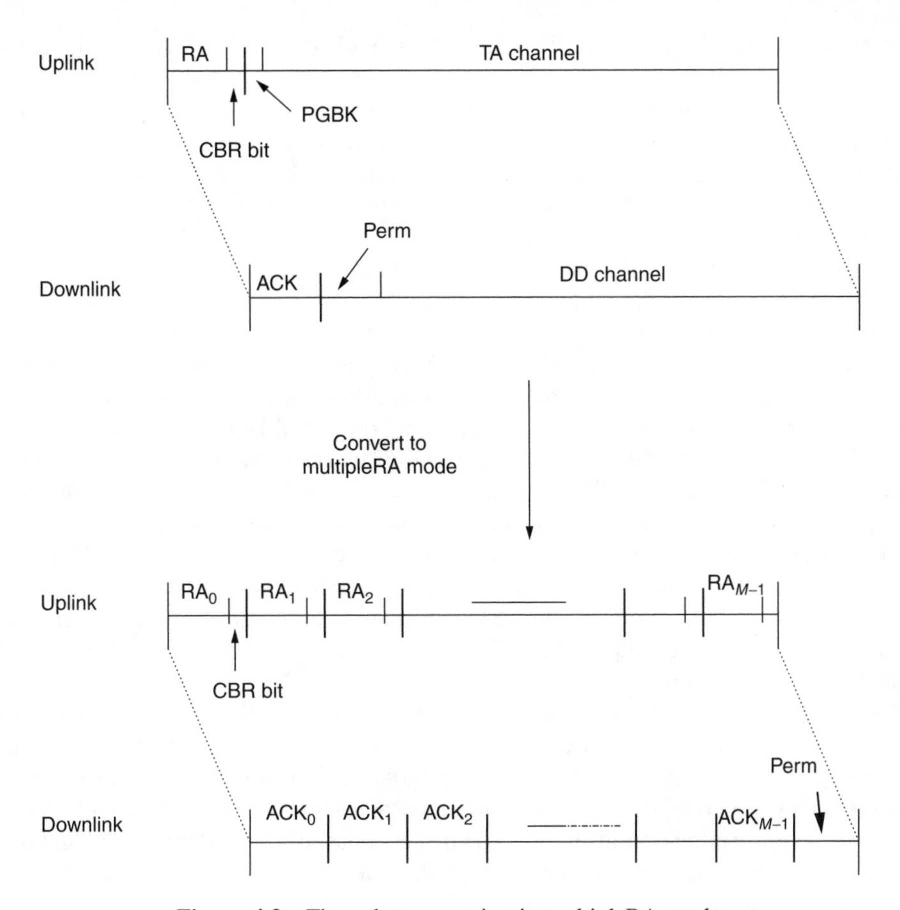

Figure 4.3 Time slot conversion in multipleRA mode.

the next time slot back to normal mode with a single RA minislot. Any remaining MSs with unsuccessful requests retransmit their requests using their corresponding q_r.

4.8 PERFORMANCE ANALYSIS

Our main design goal is to reduce channel access delay while maintaining reasonable overheads. Like in most DAMA protocols, the actual data transmission (in TA and DD slot) is collision-free. All collisions occur during the request phase. We focus on strategies that can efficiently request access without imposing severe overheads on the system. We analyze the performance of our ARCMA protocol by examining the additional feature that we introduced.

By implementing the adaptive RA channel, we improved channel utilization by exploiting unused TA slots. While avoiding the waste of valuable transmission time, this strategy also reduces contention in the RA channel. In the network with active MS, empty request queues are caused by heavy traffic where the RA channel is saturated with transmission requests (causing collisions). Therefore, it makes sense to relieve contention in the RA channel by switching to multipleRA mode. The following downlink channel has to be similarly converted to multiple ACK slots; therefore, no downlink data packets are transmitted. This imposes a single slot delay to the broadcasting of packets (downlink) to the mobiles. This is a small overhead compared to the otherwise idle uplink and downlink channel.

In our implementation, all requests are sent immediately when the channel first switches to multipleRA mode. In this mode, the number of mobile requests in each RA channel is reduced. In normal mode, five MSs have to share a single RA channel. In multipleRA mode, the maximum number of MSs per RA channel is two. Since the probability of packets arriving is assumed to be the same for all MSs, the probability of a collision, in which two or more requests are made in the same time slot, is greater in an RA channel that handles more MSs. Therefore, the probability of collisions in multipleRA mode is also reduced. The number of MSs handled by each channel is reduced by the number of RA minislots R. If the multipleRA mode does not produce any successful requests, MSs retransmit their requests (still in multipleRA mode) using their original q_r. This allows each mobile to utilize the BEB algorithm, but in a channel with lesser contention. This enhances our protocol's efficiency by reducing channel access delay (due to collisions).

Table 4.1 shows an example illustrating the number of mobiles handled by each RA channel in normal and multipleRA mode.

CBR traffic is specially handled because of its periodic characteristic that produces benefits in two areas. First, CBR traffic is transmitted with minimum delay as a result of the request-free access and transmission priority. This feature is essential since CBR traffic is delay-sensitive. Second, since no requests are needed for CBR packets (except for initial setup), contention in the RA channel is reduced. Depending on the number of mobiles with CBR traffic, this scheme can produce significant improvement to the overall system performance.

In the initial transmission request for the CBR traffic, the mobile has to send additional data representing the service type (i.e., CBR or not). Since only a single bit is required, this addition does not pose any significant overhead to the overall transmission packet.

Table 4.1 An example illustrating the number of mobiles handled by each RA channel in normal and multipleRA mode

Mobile ID	Normal mode		MultipleRA mode	
	Channel Number (single channel)	Number of mobiles per RA channel	Channel Number ($R = 3$)	Number of mobiles per RA channel
1	1	5	1	2
2	1	5	2	2
3	1	5	3	1
4	1	5	1	2
5	1	5	2	2

Note: Number of active mobiles, $M = 5$; Number of converted RA channels, $R = 3$.

Slotted ALOHA was selected as the random access protocol in the RA channel. The BEB algorithm was used to provide stability to the protocol. Such schemes reduce access delay by reducing consecutive collision in the RA channel. In addition, the multipleRA mode provides an additional layer of control for reducing collisions. In situations in which the random access protocol is unable to produce a successful request, the adaptive channel access strategy coupled with the BEB algorithm significantly reduces the collision probability in the request channel.

We summarize the relevant features of ARCMA protocol.

- *Efficient channel utilization*: Schemes such as the adaptive RA channel, the special handling of CBR traffic, and the piggyback strategy significantly improve channel utilization.
- *Slot-by-slot transmission*: MS receives ACK to transmit request almost immediately on a slot-by-slot basis. When collision occurs, MSs are quickly aware of their failed request and may retransmit in the next time slot. For a protocol that transmits on a frame-by-frame (by periods) basis, the requesting MS has to wait until the next frame before receiving any acknowledgment. A frame usually has the length (in bits) of multiple time slots. This causes delay that can be critical in a delay-sensitive service. In addition, there can be empty slots within that frame that could have been used for retransmission.
- *Transparency to AAL*: To reduce the integration complexity between wired and wireless networks, a protocol must provide seamless inter-networking such that the ATM Adaptation-Layer (AAL) is not involved. ARCMA protocol is essentially self-contained within its own network layer. The strategy does not involve the AAL.
- *Small RA packet*: In ARCMA implementation, we use a single byte (256 mobiles) request in the RA slot. Therefore, the RA slot is just a fraction of an ATM packet (53 bytes). Collision in the RA channel only wastes a small amount of the scarce wireless spectrum.
- *Preserved packet order*: Since all packets are queued in the mobile's buffer and sent sequentially on a slot-by-slot basis, the packet order is preserved. No complex reordering scheme is required at the receiving end.
- *Multiple uplink/downlink channels*: In our discussion, we assume a single uplink and downlink channel. In actual implementation, there can be multiple uplink and downlink frequencies.

4.9 PERFORMANCE MEASURES

The performance measures are the Channel Throughput (TP_C) and the Average Transmission Delay (D_{AVG}). The Average Queue Length (L_{AVG}) of the mobile's buffer illustrates the effects on CBR traffic. The performance parameters are defined as follows:

1. *Channel Throughput (TP_C)*: TP_C is defined as the ratio of the total number of transmitted packets and the total number of time slots. That is, $TP_C = P_T/T_{TL}$, where P_T is denoted as the total number of transmitted packets, and T_{TL} is denoted as the total

number of time slots. The TP_C is measured as the number of packets transmitted per time slot.

2. *Average Transmission Delay (D_{AVG})*: D_{AVG} is defined as the ratio of the total packet transmission delay and the number of active mobiles. Hence, $D_{AVG} = D_{TL}/M$, where D_{TL} is the total packet transmission delay and M is the number of active mobiles. D_{TL} is the sum of each packet transmission delay in every active mobile. Each delay is defined as the time (number of time slots) taken, when a packet first arrives at the mobile's buffer to the time the packet reaches the BS. D_{AVG} is measured by the number of time slots.
3. *Average Queue Length (L_{AVG})*: L_{AVG} is defined as the ratio of the total number of packets in all the mobiles' buffer and the number of active mobiles. Thus, $L_{AVG} = L_{TL}/M$, where L_{TL} is the total number of packets in all the mobiles' buffer, and M is the number of active mobiles. L_{AVG} is measured by the number of packets.

Protocol design goal is to reduce D_{AVG} while maintaining a reasonable TP_C.

ARCMA protocol offers better performance in terms of channel throughput and average delay under most traffic conditions. It provides better overall channel utilization by reducing contention in the RA channels. Depending on the delay tolerance of the traffic, ARCMA can achieve very high TP_C. Future high-speed cellular networks (e.g., picocell) may provide a higher delay (in time slots) tolerance enabling throughput of over 90% under suitable traffic conditions.

ARCMA protocol is designed to efficiently share the limited spectral resources of a wireless network. With the proliferation of multimedia portables, support for integrated multimedia traffic is increasingly important. In addition to the limited wireless bandwidth, new protocols are required to support real-time delay-sensitive traffic. ARCMA protocol is designed to handle some of these requirements in the MAC sublayer. There are few wireless protocols that can satisfy the high bandwidth and low Bit Error Rate (BER) of ATM networks in the wireless environment. Most of them do not provide support for the requirements of different ATM service types. The ARCMA scheme provides better support for delay-sensitive CBR traffic by prioritizing the transmission scheduling policy. In addition, ARCMA improves channel utilization by reducing collision in the request subchannel. ARCMA protocol provides request-free transmission for CBR and bursty traffic (within the same burst). An adaptive request channel can increase the request (without collision) probability by exploiting idle TA channels.

ARCMA performs better than DQRUMA regardless of the traffic load. Under heavy traffic, ARCMA protocol is capable of producing significantly higher channel throughput than DQRUMA. The worst traffic scenario for ARCMA protocol is nonbursty (single packet burst) traffic. Every packet arrival requires transmission request, causing heavy collisions in the RA channel. Conversely, ARCMA performs extremely well with bursty traffic (e.g., VBR) capable of achieving over 85% channel throughput with limited transmission delay. The CBR extension enables ARCMA to satisfy the delay-sensitive CBR traffic while reducing collisions in the RA channel. This result justifies the added complexity and overhead for CBR support.

Although the ARCMA protocol does not provide direct support to the other time-sensitive traffic (e.g., VBR), the strategies implemented in ARCMA protocol significantly reduce contention in the RA channel, allowing such traffic to transmit with less delay. ARCMA provides an efficient DAMA that is practical for implementation in a Wireless ATM (WATM) network. It brings us a step closer to designing a complete protocol suite that could be used in the wireless integration of ATM networks.

ARCMA protocol can be extended to provide direct support for other ATM services such as VBR and ABR traffic. Access delay can be further reduced if there exists a mechanism to specifically handle VBR or ABR mobiles. Such a mechanism alleviates the need for retransmitting requests packets through the RA channel. ARCMA protocol does not include services for network management. To provide a complete MAC sublayer support, we need to include services such as call admission and cell handoff.

4.10 SUMMARY

RS-ISMA is a wireless access protocol designed for wireless multimedia communications and implemented in the BRAIN indoor-LAN prototype. In addition, a compact RF module composed of flat antennas and an MMIC was employed for each remote station and AP. The use of large capacity FPGA decreased the number of signal processing boards. System parameters such as the packet format were optimized for IP datagram transport to support all applications based on IP. The function of NACK sensing was added to RS-ISMA to ensure an efficient and smooth wireless multicast in a multiple-access environment.

The HAMAC protocol uses a super frame that is divided into two frames, the downlink frame and the uplink frame. The length of the frames can vary depending on the bandwidth demand. The downlink frame is used by the BS to broadcast the frame configuration information, the connection setup, the allocation information, the request information, and the data to all mobile devices. The information and the data can be broadcast using a single burst because only the BS controls the downlink. Mobile devices can filter out irrelevant information upon receiving them. The first segment of the downlink frame is used for control signaling needed for the frame configuration to be known by all mobile devices before starting the reception and the transmission.

ARCMA implements a dynamic RA channel in which an entire uplink channel can be converted into multiple RA channels. This conversion is done when the Request Table is empty, which in most cases indicates heavy collisions in the request channel. ARCMA uses an algorithm that takes advantage of the random access scheme in the RA channel. We use the slotted ALOHA with BEB as the random access protocol for ARCMA.

The request is made in the RA channel (RA minislot). The request data packet contains the mobile's b-bit Access ID assigned during setup. In ARCMA protocol, in addition to the Access ID, the request packet also includes the type of service being requested. The protocol provides additional support for periodic traffic (i.e., CBR). Since traffic can be either CBR or non-CBR, only a single bit is required to identify the service type. This bit is transmitted together with the request packet in the RA channel. DQRUMA provides no distinction between requests of different service types. The distinction provided in ARCMA is used by the BS to assign priority to CBR traffic.

PROBLEMS TO CHAPTER 4

Wireless protocols

Learning objectives

After completing this chapter you are able to

- demonstrate an understanding of different wireless protocols.
- explain a MAC protocol for wireless LAN.
- explain implementation of BRAIN architecture.
- explain the HAMAC protocol.
- demonstrate an understanding of demand assignment multiple access protocols.
- explain the role of a Request Table in ARCMA.
- explain implementation of multiple RA channels.

Practice problems

4.1: What is the role of network and native service access points?
4.2: What is the RS-ISMA?
4.3: What are the functions of HAMAC protocol?
4.4: How is transmission performed in ARCMA?
4.5: What is the role of a Request Table?
4.6: How is dynamic RA channel implemented?

Practice problems solutions

4.1: A MAC protocol for a wireless LAN provides two types of data transfer SAP: network and native. The network SAP offers an access to a legacy network protocols (e.g., IP). The native SAP provides an extended service interface that may be used by custom network protocols or user applications capable of fully exploiting the protocol specific QoS parameters within the cell service area.

4.2: RS-ISMA is a wireless access protocol designed for wireless multimedia communications and implemented in the BRAIN indoor-LAN prototype. RS-ISMA is a wireless MAC protocol, which is an integration of reservation-based ISMA and slotted ISMA, and is basically a combination of random access protocol and polling protocol. During the reservation step, an ST transmits a short frame to make a reservation under a random access scheme. In the information transmission step, either an isochronous or an asynchronous polling scheme is used for information transmission depending on the QoS requirements.

4.3: The HAMAC protocol integrates fixed assignment TDMA protocols, reservation-based protocols, and contention-based protocols into a wireless network, simultaneously and efficiently supporting various classes of traffic such as CBR, VBR, and ABR traffic. The HAMAC protocol uses a preservation slot technique to minimize the packet contention overhead in PRMA protocols, while retaining most isochronous service features of TDMA protocols to serve voice and CBR traffic streams.

4.4: ARCMA is a DAMA protocol with dynamic bandwidth allocation. This scheme is designed to function in a cell-based wireless network with many MSs communicating with the BS of their particular cell. Transmissions are done on a slot-by-slot basis without any frames. Each slot is divided into a TA slot and an RA minislot. The RA channel in ARCMA is capable of carrying additional information for different classes of ATM service (e.g., CBR, VBR, etc.). This additional information is used by the BS to provide better QoS support for different classes of traffic. Transmission from CBR traffic may reserve an incremental series of slots in the duration of their transmission. No further request is needed until the CBR transmission finishes.

4.5: The BS maintains a Request Table to keep track of all successful requests and assigns permission to mobiles for transmission at different time slots. In ARCMA protocol, the BS inspects the service class of a request and gives transmission priority to delay-sensitive data (e.g., CBR). A piggyback bit is used in the uplink channel to reduce contention in the RA channel. This is especially beneficial for bursty traffic.

4.6: ARCMA implements a dynamic RA channel in which an entire uplink channel can be converted into multiple RA channels. This conversion is done when the Request Table is empty, which in most cases indicates heavy collisions in the request channel. ARCMA uses an algorithm that takes advantage of the random access scheme in the RA channel. We use the slotted ALOHA with BEB as the random access protocol for ARCMA.

ARCMA improves the spectral efficiency by reducing collisions in the RA channel while improving support for the various classes of ATM services.

5

Protocols for wireless applications

Wireless data networks present a more constrained communication environment compared to wired networks. Because of fundamental limitations of power, available spectrum, and mobility, wireless data networks tend to have less bandwidth than traditional networks, more latency than traditional networks, less connection stability than other network technologies, and less predictable availability.

Mobile devices have a unique set of features that must be exposed in the Web, in order to enable the creation of advanced telephony services that include location-based services, intelligent network functionality, including integration into the voice network, and voice/data integration.

The Wireless Application Protocol (WAP) architecture provides a scalable and extensible environment for application development for mobile communication devices. The WAP protocol stack has a layered design, and each layer is accessible by the layers above, and by other services and applications. The WAP layered architecture enables other services and applications to use the features of the WAP stack through a set of well-defined interfaces. External applications can access the session, transaction, security, and transport layers directly.

5.1 WIRELESS APPLICATIONS AND DEVICES

Providing Internet and World Wide Web (WWW) services on a wireless data network presents many challenges because most of the technology developed for the Internet has been designed for desktop and larger computers that support medium to high bandwidth connectivity over generally reliable data networks.

Mobile and wireless devices are usually handheld devices, and accessing the WWW presents a more constrained computing environment compared to desktop computers because of fundamental limitations of power and form factor. Mass-market handheld wireless devices tend to have

- less powerful CPUs (Central Processor Units)
- less memory [both ROM (Read Only Memory) and RAM (Random Access Memory)]

- restricted power consumption
- smaller displays
- different input devices (e.g., a phone keypad, voice input, etc.).

Wireless data networks also present a more constrained communication environment compared to wired networks. Because of fundamental limitations of power, available spectrum, and mobility, wireless data networks tend to have

- less bandwidth than traditional networks;
- more latency than traditional networks;
- less connection stability than other network technologies; and
- less predictable availability.

Mobile networks are growing in complexity and the cost of providing new value-added services to wireless users is increasing. To meet the requirements of mobile network operators, solutions must be

- interoperable – terminals from different manufacturers communicate with services in the mobile network;
- scalable – mobile network operators should be able to scale services to customer needs;
- efficient – provide quality of service suited to the behavior and characteristics of the mobile network; provide for maximum number of users for a given network configuration;
- reliable – provide a consistent and predictable platform for deploying services;
- secure – enable services to be extended over potentially unprotected mobile networks while still preserving the integrity of user data; protect the devices and services from security problems such as denial of service.

The WAP Forum is an industry group dedicated to the goal of enabling sophisticated telephony and information services on handheld wireless devices such as mobile telephones, pagers, Personal Digital Assistants (PDAs), and other Wireless Terminals (WTs). Recognizing the value and utility of the WWW architecture, the WAP Forum has chosen to align certain components of its technology very tightly with the Internet and the WWW. The WAP specifications extend and leverage mobile networking technologies (such as digital data networking standards) and Internet technologies, such as IP, Hypertext Transfer Protocol (HTTP), Extensible Markup Language (XML), Uniform Resource Locators (URLs), scripting, and other content formats.

The WAP Forum drafted a global wireless protocol specification for all wireless networks and contributes it to the industry and standards bodies. WAP enables manufacturers, network operators, content providers, and application developers to offer compatible products and secure services on all devices and networks, resulting in greater economies of scale and universal access to information.

The objectives of the WAP Forum are

- to bring Internet content and advanced data services to digital cellular phones and other WTs;

- to create a global wireless protocol specification that works across different wireless network technologies;
- to enable the creation of content and applications that scale across a very wide range of wireless bearer networks and wireless device types;
- to embrace and extend existing standards and technology wherever appropriate.

To bring Internet and WWW technologies to digital cellular phones and other WTs, that is, adapting the Web architecture to the wireless environment, and to enable the delivery of sophisticated information and services to mobile WTs requires working toward a unified information space, common standards, and technologies.

Wireless network bearers operate under several fundamental constraints, which place restrictions on the type of protocols and applications offered over the network:

- *Power consumption*: As bandwidth increases, power consumption increases. In a mobile device, this reduces battery life.
- *Cellular network economics*: Mobile networks are typically based on a cellular architecture. Cells are a resource shared by all mobile terminals in a geographic area and typically have a fixed amount of bandwidth to be shared among all users. This characteristic rewards efficient use of bandwidth, as a means of reducing the overall cost of the network infrastructure.
- *Latency*: The mobile wireless environment is characterized by a very wide range of network latency, ranging from less than a second round-trip communication time to many tens of seconds. In addition, network latency can be highly variable, depending on the current radio transmission characteristics (e.g., in a tunnel or off network) and the network loading in a particular area. Latency is further increased by routing, error correction, and congestion avoidance characteristics of a particular network.
- *Bandwidth*: The mobile wireless environment is characterized by a very wide range of network characteristics and typically has far less bandwidth available than a wireline environment. In addition, the economics of the wireless environment encourage the conservation of bandwidth to achieve greater density of subscribers.

Wireless devices operate under a set of physical limitations, imposed by their mobility and form factor:

- *Limited power*: Any personal or handheld mobile device will have a very limited power reserve, owing to existing battery technology. This reduces available computational resources, transmission bandwidth, and so on.
- *Size*: Many mobile wireless devices are very small (handheld).

Mobile wireless devices are characterized by a different set of user interface constraints than that of a personal computer. To enable a consistent application-programming model, a very wide range of content scalability is required. In practice, a significant amount of the WWW content is unsuitable for use on handheld wireless devices. The problems include the following:

- *Output scalability*: Existing content is designed for viewing on PC (Personal Computer) screens, whereas mobile devices have a wide range of visual display sizes, formatting and other characteristics that include voice-only output.

- *Input scalability*: Mobile devices feature a wide range of input models, including numeric keypad, very few or no programmable soft keys, and so on, and voice-only input.

Many wireless devices, for example, cellular phones and pagers, are consumer devices. These devices are used in a wide variety of environments and in a wide range of scenarios. The examples include the following:

- *Simple user interfaces*: Many mobile devices, in particular, cellular telephones, are mass-market consumer-oriented devices. Their user interface must be extremely simple and easy to use.
- *Single-purpose devices*: The goal and purpose of most mobile devices is very focused (e.g., voice communication). This is in contrast with the general-purpose tool-oriented nature of a personal computer. This motivates a very specific set-of-use cases, with very simple and focused behavior, for example, placing a voice call.
- *Hands-free, heads-up operation*: Many mobile devices are used in environments in which the user should not be unnecessarily distracted (e.g., driving and talking).

The World Wide Web Consortium (W3C) is leading and participating in the continuing development of the Web and its standards. The new generation of Web technologies is intended to enhance the users' and publishers' control over the presentation of the information [e.g., through Cascading Style Sheets (CSS)], over the management of information [e.g., through Resource Description Framework (RDF)], and over its distribution [e.g., through P3P (Platform for Privacy Preferences Project)] on the basis of technologies that structure and distribute data as objects, such as XML and HTTP-NG (Network Group). These technologies will be described later in the text.

A new generation of Hypertext Markup Language (HTML) is based on XML and includes features that make it more efficient for mobile use. The other XML applications such as the Wireless Markup Language (WML) and the Synchronized Multimedia Integration Language (SMIL) have components where mobile access has an impact.

A Scalable Vector Graphics (SVG) format, which is written as a modular XML tagset and is usable as an XML name space, can be widely implemented in browsers and authoring tools and is suitable for widespread adoption by the content authoring community as a replacement for many uses of raster graphics. In simple cases such as in-line graphics, it should be possible to hand the author the SVG format, and it should also be possible to cut and paste SVG graphical objects between documents and to preserve their appearance, linking behavior, and style. The graphics in Web documents are smaller, faster, more interactive, and displayable on a wider range of device resolutions from small mobile devices through office computer monitors to high-resolution printers.

In the presentation model for the new generation of Web technologies, the formatting of a document is conducted through the use of a style sheet. This is a separate document that allows authors and users to attach style (e.g., fonts, spacing, and aural cues) to structured documents (e.g., HTML documents and XML applications). By separating the presentation style of documents from the content of documents, Cascading Style Sheets Level 2 (CSS2) and Extensible Stylesheet Language (XSL) simplifies Web and XML authoring and site maintenance. Local processing of a document might in the future also be conducted using a similar technology called *action sheets*. Style sheets can have media-specific properties, which makes them a possible candidate for use with mobile devices.

The Document Object Model is a platform- and language-neutral interface that allows programs and scripts to dynamically access and update the content, structure, and style of documents. The Document Object Model provides a standard set of objects for representing HTML and XML documents, a standard model of how these objects can be combined, and a standard interface for accessing and manipulating them.

The purpose of the HTTP-NG activity is to design, implement, and test a new architecture for the HTTP protocol on the basis of a simple, extensible, distributed object-oriented model. This includes a protocol for the management of the network connections, a protocol for transmitting messages between systems, a set of methods, interfaces, and objects that demonstrate a classical Web browsing case, as an example of what is possible with the new protocol and a test bed to test the implementation.

Accessibility for people with disabilities is relevant for mobile wireless devices as this is a potentially large marketplace (over 10% of the population), and in some cases accessibility is required (e.g., for sales in the United States, under Section 255 of the US Telecommunications Act). In addition, functions, such as speech input or output, required to accommodate different kinds of disability have carry-over benefits for nondisabled users of mobile devices, who may be using the devices in hands-free or eyes-free situations.

W3C's Web Accessibility Initiative (WAI), in coordination with other organizations, is addressing Web accessibility through several areas of work and related technology and guidelines to mobile wireless devices. In the area of technology, WAI works with W3C Working Groups developing technologies that can facilitate accessibility, such as HTML, CSS, SMIL, and SVG. In the area of guidelines, WAI is developing guidelines for accessible page authoring, user agents, and authoring tools and is coordinating with the development of guidelines by the Mobile Access Interest Group.

The correct representation of characters is an issue in all formats of writing, not just the Latin alphabet. The aim of this activity is for the WWW to live up to its name, and the W3C continues work on the internationalization of the Web with the aim of ensuring that the necessary features are included in W3C protocols and data format recommendations. The general goal of W3C's work on internationalization is to ensure that W3C's formats and protocols are usable worldwide in all languages and writing systems.

Establishing trust in the new medium of the Web involves both social and technical issues. Trust is established through a complex and ill-understood social mechanism including relationships, social norms, laws, regulations, traditions, and track records.

There is a core of technical issues that are required in any system that is to be trusted:

- The ability to make statements that have agreed-upon meanings. The W3C Metadata Activity provides a means to create machine-readable statements.
- The ability to know who made the statement and to be assured that the statement is really theirs. The W3C Digital Signature Initiative provides a mechanism for signing metadata in order to establish who is making the machine-readable statement.
- The ability to establish rules that permit actions to be taken, based on the statements and a relationship to those who made the statements. The Platform for Internet Content

Selection (PICS) rules specification allows rules to be written down so that they can be understood by machines and exchanged by users.

- The ability to negotiate binding terms and conditions. The Joint Electronic Payment Initiative (JEPI) project created the Protocol Extension Protocol (PEP) to provide for negotiation on the Web. Negotiation is also at the core of the Platform for Privacy Preferences Project (P3P).
- Electronic commerce markup and payment: The W3C has two working groups in this field, on markup for electronic commerce and for payment initiation.

The WAP Forum's exclusive focus is mobile wireless technologies. The goal of WAP is to create recommendations and specifications that support the creation of advanced services on wireless devices with particular emphasis on the mobile telephone. The WAP Forum is creating recommendations and technologies, which enable these services on all mobile devices and on all networks.

The WAP Forum has undertaken a variety of technical specification work relevant to the W3C/WAP Forum collaborative efforts. All these efforts relate to the use of World Wide Web technology on mobile devices, and in ensuring that the quality of these services is sufficient for mass deployment.

WAP is focused on enabling the interconnection of the Web and WTs. Significant focus has been given to mobile telephones and pagers, but all technology has been developed with broader applicability in mind. The goal of WAP is to enable an extremely wide range of WTs that range from mass-market mobile telephones and pagers to more powerful devices to enjoy the benefits of Web technology and interconnection.

Mobile devices have a unique set of features, which must be exposed in the Web, in order to enable the creation of advanced telephony services, and include

- location-based services;
- intelligent network functionality, including integration into the voice network;
- voice/data integration.

The WAP Forum is working to increase the bandwidth efficiency of Web technology to make it more applicable to the wireless environment. WAP Forum work includes the following:

- Smart Web proxies – proxies capable of performing intelligent transformation of protocols and content, enabling more efficient use of the network, adaptation to device characteristics, and adaptation to network characteristics.
- Efficient content encoding – bandwidth efficient encodings of standard Web data formats such as XML.
- Efficient protocols – bandwidth efficient adaptations of standard Web protocols such as HTTP.

The WAP Forum is working to improve the behavior of Web technology due to high network latencies, and in particular, is focusing on the problems of

- tuning network protocols to be adaptive and efficient given wide ranging latencies;
- creating Web applications that are resilient to either high latency environments or highly variable latency situations.

WAP Forum work in this area includes the following:

- User agent state management
- Protocol design (e.g., session state, fast session resumption, etc.).

Mobile wireless devices are characterized by a different set of user interface constraints than a personal computer. The WAP Forum work in this area includes the following:

- Content adaptation – mechanisms allowing a Web application to adapt gracefully to the characteristics of the device (beyond the HTTP/1.1 content negotiation model).
- User interface scalability content formats – for example, markup and display languages that are suitable to impoverished devices, but which scale well to more sophisticated devices.

In the area of Web technologies, the focus of the WAP Forum and the W3C directly overlaps in the areas of intelligent proxies and protocol design, in XML applications, and in content adaptation, for example, through the use of vector graphics and style sheets. The cooperation may also occur in the area of electronic payment in which the work of both groups has the potential to overlap.

Instead of developing diverging solutions, it is the intent of both groups to find common solutions that will address mobile requirements. In the area of Web technology, the goals overlap, especially in the long run, allowing significant cooperation and shared development. To avoid fragmentation of the Web standards, the groups cooperate and focus on achieving the seamless integration of mobile devices into the Web.

5.2 MOBILE ACCESS

The idea of access to the Web from any place and at any time is fast becoming a reality. Web information and services are becoming accessible from a wide range of mobile devices, from cellular phones, pagers, and in-car computers to palmtop computers and other small mobile devices. Many such devices are characterized by small screens, limited keyboard, low bandwidth connection, and small memory.

Mobile devices need special consideration when it comes to using Web information. Their displays are generally much smaller than a conventional computer screen and are capable of showing only a small amount of text. On a cellular phone, for example, there may be only enough space for three or four rows of text. Palmtop pocket-sized computers have screens smaller than a PC or a laptop, but large enough to read e-mail (electronic mail) and documents with a small amount of text. Mobile devices have limited memory and processing speeds, and these considerations also need to be taken into account.

Mobile devices may not use all the HTML tags of a normal Web page. Given that mobile devices are different in their capabilities from ordinary PCs, what are the repercussions for markup? Because of the constraints explained above, mobile devices are unlikely to be able to use exactly the same markup as a normal page for a PC. Instead, they will use a subset of HTML tags. The expectation is that different devices will make use of different modules of Extensible HTML (XHTML); similarly, they will support different modules of style sheets. For example, one mobile device may use the basic XHTML text module and the style sheet voice module. Another device with a large screen may also allow the XHTML tables module.

How can a device tell the server about its capabilities? The question is, given the needs of the various devices accessing the Web, how can the server know about the capabilities of individual devices? How can it know that a mobile phone with a very small screen is requesting a Web page, rather than a pocket-sized computer asking for the same information? The idea is to store data about each device, and also the preferences of its user, as a device profile. The device profiles are stored as a kind of relational database located on a Web server. W3C is working jointly with the WAP Forum writing the database model and its fields. This work has led to the Composite Capability/Preference Profiles (CC/PP).

A CC/PP is a collection of information, which describes the capabilities, hardware, system software, and applications used to access the Web, and the particular preferences of the users themselves. Information may include the preferred language, sound on/off, images on/off, class of device (phone, PC, printer, etc.), screen size, available bandwidth, version of HTML supported, and so on.

The location of the device profile is sent with a request for a Web page. When a device makes a request over the Web for a Web page, a pointer to the device profile is appended to the request. In the case of a mobile phone, the phone requests a Uniform Resource Identifier (URI) in the usual way and sends a pointer in the form of a second URI to indicate where its device profile can be found.

The pointer URI goes straight to the CC/PP database. CC/PP is written in RDF, W3C's language for modeling metadata, descriptive information about items on the Web. In RDF, the information encoded is always linked to Web addresses. This means that by sending a URI for the device profile, all kinds of data about that device immediately becomes available.

On the basis of the device profile, a Web server can choose the right content since the device profile is known and the Web information required is understood. XHTML is designed as a series of modules associated with different functionality: text, tables, forms, images, and so on. CSS and SMIL specifications have the same modular construction. If a content provider wants information to be available for different devices, different versions of that content can be generated, for example, by using only the text modules, or by using full graphics with scripting. Thus, in its document profile, the document specifies the expected capabilities of the browser in terms of XHTML support, and style sheet support. During the process of matching, the document profile is compared with the device profile, the best fit between the two is discovered, and a suitable document is generated or the best fitting variant is selected.

5.3 XML PROTOCOL

Structured data such as spreadsheets, address books, configuration parameters, financial transactions, technical drawings, and so on, are often stored on a disk, for which either a binary format or a text format can be used. The latter allows the user, if necessary, to look at the data without the program that produced it. XML is a set of rules, guidelines, and conventions for designing text formats for such data, in a way that produces files that are easy to generate and read by a computer, that are unambiguous and that avoid common pitfalls such as lack of extensibility, lack of support for internationalization/localization, and platform-dependency.

XML makes use of tags, but while HTML specifies what each tag and attribute means (and often how the text between them will look in a browser), XML uses the tags only to delimit pieces of data and leaves the interpretation of the data completely to the application that reads it.

XML files are text files because that allows experts, such as programmers, to more easily debug applications, and in emergencies they can use a simple text editor to fix a broken XML file. But the rules for XML files are much stricter than for HTML. A forgotten tag, or an attribute without quotes makes the file unusable, while in HTML such practice is often explicitly allowed, or at least tolerated. In the official XML specification, the applications are not allowed to try to second-guess the creator of a broken XML file; if the file is broken, an application has to stop right there and issue an error message.

Since XML is a text format, and it uses tags to delimit the data, XML files are nearly always larger than comparable binary formats. That was a conscious decision by the XML developers. Communication protocols such as modem protocols and HTTP/1.1 (the core protocol of the Web) can compress data, thus saving bandwidth as effectively as a binary format.

Data transport is as central to modern computing as data storage and display in the networked, decentralized, and distributed environment of the Internet and Web. Following the adoption of XML for data processing, the challenge is for both sides of a session to agree on an application-layer transfer protocol, whether between software programs, between machines, or between organizations. Even though it accounts for most Web surfing, interactive browsing by human operating user agents can accomplish only so much alone.

To automate negotiations and to stimulate the Web's growth, standardized application-to-application messaging is required. The search is on for common ground that can meet the heavy weight, commercial demands of business-to-business electronic commerce systems and at the same time satisfy aesthetic requirements for a lightweight, simple network protocol for distributed applications.

W3C's XML Protocol Activity addresses these needs. Its XML Protocol Working Group is chartered to design the following:

- An envelope to encapsulate XML data for transfer in an interoperable manner that allows for distributed extensibility, evolvability, as well as intermediaries such as proxies, caches, and gateways;
- In cooperation with the Internet Engineering Task Force (IETF), an operating system-neutral convention for the content of the envelope when used for Remote Procedure Call (RPC) applications;
- A mechanism to serialize data based on XML Schema data types; and
- In cooperation with the IETF, a nonexclusive mechanism layered on HTTP transport.

W3C provides the platform for discussion and for planning and creation of an XML Protocol Recommendation. Through rigorous examination of the various XML protocols in development or those already deployed, the XML Protocol Working Group is creating an open specification for an interoperable protocol for use by all interested parties. Working together with the IETF, W3C also cooperates on efforts to build on HTTP. The XML Protocol Working Group also has contact with members of the Transport, Routing, and Packaging project. The XML Protocol Working Group also participates in the XML Coordination Group to assure coordination with related XML efforts.

5.4 DATA ENCAPSULATION AND EVOLVABILITY

For two peers to communicate in a distributed environment, they must first agree on a unit of communication. The XML Protocol Working Group defines an encapsulation language that allows for applications to independently introduce extensions and new features. The following requirements for extensions and features must be met:

- They are or can be orthogonal to other extensions.
- They can be deployed automatically and dynamically across the Web with no prior coordination and no central authority.
- The sender can require that the recipient either obeys the semantics defined by an extension or aborts the processing of the message.

The Extensible Protocol (XP) specification must define the concept of an envelope or outermost syntactical construct or structure within which all other syntactical elements of the message must be enclosed. The envelope must be described with XML Schema.

The XP specification must also define a processing model that defines what it means to properly process an XP envelope or to produce a fault. This processing model must be independent of any extensions carried within the envelope. The processing model must apply equally to intermediaries as well as to ultimate destinations of an XP envelope.

The XP specification must define a mechanism or mechanisms that allow applications to submit application-specific content or information for delivery by XP. In forming the standard for the mechanisms, the XP specification may consider support for

- carrying application-specific payloads inside the XP envelope,
- referring to application-specific payloads outside the XP envelope,
- carrying nested XP envelopes as application-specific data within the XP envelope,
- referring to XP envelopes as application-specific data outside the XP envelope,
- extending the message by extension of the XP envelope itself.

To manage the mechanisms, the XP specification must define a set of directives that unambiguously indicate to an XP processor which extensions are optional and which are mandatory so that it can

- process all the extensions in an XP envelope or fail,
- process a subset of the extensions in an XP envelope or fail.

In both the cases mentioned above, the XP processor must fail in a standard and predictable fashion.

The XP specification must define the concept of protocol evolution and define a mechanism or mechanisms for identifying XP revisions. This mechanism or mechanisms must ensure that given two XP messages it should be possible, by simple inspection of the messages, to determine if they are compatible. The specification must define the concepts of backwards compatible and backwards incompatible evolution. Furthermore, the XP envelope must support both optional and mandatory extensibility of applications using the XP envelope.

The XP specification must define a means to convey error information as a fault. The capability of XP carrying a fault message must not depend on any particular protocol binding. The XP specification must define a mechanism or mechanisms to allow the transfer of status information within an XP message without resorting to the use of XP fault messages or without depending on any particular interaction model.

Intermediaries are essential parts of building distributed systems that scale to the Web. Intermediaries can act in different capacities ranging from proxies, caches, store-and-forward hops to gateways. Experience from HTTP and other protocols has shown that intermediaries cannot be implicitly defined but must be an explicit part of the message path model for any data encapsulation language. Therefore, the Working Group must ensure that the data encapsulation language supports composability both in the vertical (within a peer) as well as in the horizontal (between peers) dimension.

Intermediaries are essential parts of building distributed systems that scale on the Web. As a result, XML Protocol must support intermediaries. Because XML Protocol separates the message envelope from the transport binding, two types of intermediaries are possible – transport intermediaries and processing intermediaries.

With the introduction of XML and RDF schema languages, and the existing capabilities of object and type modeling languages such as Unified Modeling Language (UML), applications can model data at either a syntactic or a more abstract level. In order to propagate these data models in a distributed environment, it is required that data conforming to a syntactic schema can be transported directly, and that data conforming to an abstract schema can be converted to and from XML for transport.

The Working Group should propose a mechanism for serializing data representing nonsyntactic data models in a manner that maximizes the interoperability of independently developed Web applications. Furthermore, as data models change, the serialization of such data models may also change. Therefore, it is important that the data encapsulation and data representation mechanisms are designed to be orthogonal.

Examples of relationships that will have to be serialized include subordinate relationships known from attachments and manifests. Any general mechanism produced by the Working Group for serializing data models must also be able to support this particular case.

The XML Protocol data encapsulation and data representation mechanisms must be orthogonal. The XML Protocol data representation must support using XML Schema, simple and complex types. The XML Protocol data representation must be able to serialize data based on data models not directly representable by XML Schema, simple and complex types. These data models include object graphs and directed labeled graphs. It must be possible to reconstruct the original data from the data representation. Data serialized according to the XML Protocol data representation may contain references to data outside the serialization. These references must be URIs. The XML Protocol data representation must be able to encode arrays, which may be nested.

A mechanism for using HTTP transport is needed in the context of an XML Protocol. This does not mean that HTTP is the only transport mechanism that can be used for the technologies developed, nor that support for HTTP transport is mandatory. This component merely addresses the fact that HTTP transport is expected to be widely used.

Mapping onto existing application layer protocols may lead to scalability problems, security problems, and semantic complications when the application semantics defined by those protocols interfere with the semantics defined by an XML Protocol.

General transport issues were investigated by the HTTP-NG activity, which designed a general transport mechanism for handling out-of-order delivery of message streams between two peers.

The XP specification must not mandate any dependency on specific features or mechanisms provided by a particular transport protocol beyond the basic requirement that the transport protocol must have the ability to deliver the XP envelope as a unit. This requirement does not preclude a mapping or binding to a transport protocol taking advantages of such features. It is intended to ensure that the basic XP specification will be transport neutral.

The XP specification must consider the scenario in which an XP message may be routed over possibly many different transport or application protocols as it moves between intermediaries on the message path. This requirement implies it must be possible to apply many transport or application protocol bindings to the XP message without information loss from the message. The XP specification should not preclude the use of XP messaging over popular security mechanisms.

The XP specification must provide a normative description of the default binding of XP to the HTTP transport. This binding, while normative, is not to be exclusive. Any protocol binding to HTTP must respect the semantics of HTTP and should demonstrate that it can interoperate with existing HTTP applications. This requirement does not extend to the provision of normative bindings to other important Internet protocols such as Transmission Control Protocol/Internet Protocol (TCP/IP), User Datagram Protocol (UDP), and Simple Message Transfer Protocol (SMTP) in the XP specification.

Furthermore, the XP specification must provide a normative description of a binding of XP to a subset of HTTP that is compatible with pre-XP Internet browser technology.

Given below is the convention for the content of the envelope when used for RPC applications. The protocol aspects of this should be coordinated closely with the IETF.

The XML Protocol contains a convention for representing calls and replies between RPC applications. The conventions include the following:

- Unique specification of the program or object and procedure or method to be called on the basis of the URI syntax.
- The ability to specify the parameters to a call in a request message and the results of a call in the reply messages.
- Provisions for specifying errors in a reply message. Where possible, an attempt will be made to leverage any related work done by the IETF.

The RPC conventions within the XML Protocol use the Data Representation model to represent parameters to a call in the request message and results of the call in the reply message. There is straightforward mapping of the data types used in a wide variety of widely deployed programming languages and object systems.

The XML Protocol allows applications to include custom encodings for data types used for parameters and results in RPC messages. Mechanisms for automatically binding data represented in RPC messages to native constructs in a programming language are not precluded.

The XML Protocol will guarantee that RPC messages that encode parameters and results using the default encoding for the base set of data types will be valid for any conformant binding of the RPC conventions. Valid in this context means that the semantics of the call should remain identical, irrespective of the programming language or object system used by the caller or receiver. The subsections contained within are the same as the subsections of the out-of-scope section of the charter.

Direct handling of binary data

XML name spaces provide a flexible and lightweight mechanism for handling language mixing as long as those languages are expressed in XML. In contrast, there is only very rudimentary support (base-64 encodings, etc.) for including data languages expressed in binary formats. Such formats include commonly used image formats such as Portable Network Graphics (PNG), Joint Photographic Experts Group (JPEG), and so on.

Transport services are extremely important in order to actually deliver packages in an efficient and scalable manner. XML messaging proposals use existing application layer protocols such as SMTP and HTTP. The XML Protocol Working Group focuses on providing a (nonexclusive) mapping to HTTP.

Mapping onto existing application layer protocols may lead to scalability problems, security problems, and semantic complications when the application semantics defined by those protocols interfere with the semantics defined by an XML Protocol.

5.5 WIRELESS APPLICATION PROTOCOL (WAP)

The WAP architecture provides a scalable and extensible environment for application development for mobile communication devices. The WAP protocol stack has a layered design, and each layer is accessible by the layers above and by other services and applications. The WAP layered architecture enables other services and applications to use the features of the WAP stack through a set of well-defined interfaces. External applications can access the session, transaction, security, and transport layers directly. The protocol stack of the WAP architecture is shown in Figure 5.1.

The Wireless Application Environment (WAE) is a general-purpose application environment based on the combination of WWW and Mobile Telephony technologies. The WAE allows operators and service providers to build applications and services that can reach wireless platforms in an efficient and useful manner. WAE contains a microbrowser environment containing the following functionality:

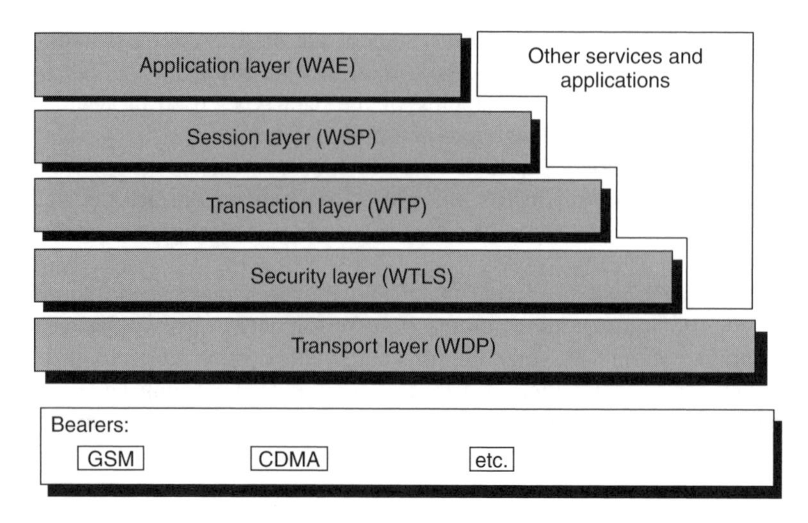

Figure 5.1 WAP architecture.

- *Wireless Markup Language (WML)*: a lightweight markup language, similar to HTML, and optimized for use in handheld mobile devices;
- *WMLScript*: a lightweight scripting language, similar to JavaScript;
- *Wireless Telephony Application (WTA)*: telephony services and programming interfaces; and
- *Content formats*: a set of well-defined data formats, including images, phone book records, and calendar information.

Wireless Session Protocol (WSP) provides the application layer of WAP with an interface for two session services. The connection-oriented service operates above the Wireless Transaction Protocol (WTP). The connectionless service operates above a secure or nonsecure Wireless Datagram Protocol (WDP). The WSP consists of services for browsing applications providing the following functionality:

- HTTP/1.1 functionality and semantics in a compact over-the-air encoding;
- Long-lived session state;
- Session suspend and resume with session migration;
- A common facility for reliable and unreliable data push; and
- Protocol feature negotiation.

The WTP runs on top of a datagram service and is suitable for implementation of Mobile Stations (MSs) as a lightweight transaction oriented protocol. WTP operates efficiently over secure or nonsecure wireless datagram networks and provides the following features:

- Three classes of transaction service: unreliable one-way requests, reliable one-way requests, and reliable two-way request-reply transactions;

- Optional user-to-user reliability – WTP user triggers the confirmation of each received message;
- Optional out-of-band data on acknowledgments;
- Protocol Description Unit (PDU) concatenation and delayed Acknowledgement to reduce the number of messages sent; and
- Asynchronous transactions.

Wireless Transport Layer Security (WTLS) is a security protocol based upon the industry standard Transport Layer Security (TLS) protocol, formerly known as Secure Sockets Layer (SSL). WTLS is used with the WAP transport protocols and is optimized for use over narrow-band communication channels. WTLS provides the following features:

- *Data integrity*: WTLS ensures that data sent between the terminal and an application server is unchanged and uncorrupted;
- *Privacy*: WTLS ensures that data transmitted between the terminal and an application server is private and cannot be understood by any intermediate parties that may have intercepted the data stream;
- *Authentication*: WTLS contains facilities to establish the authenticity of the terminal and application server;
- *Denial of service protection*: WTLS detects and rejects data that is replayed or not successfully verified. WTLS protects the upper protocol layers.

WTLS can also be used for secure communication between terminals, for authentication of electronic business card exchange. The applications can selectively enable or disable WTLS features depending on their security requirements and the characteristics of the underlying network.

WDP is the transport layer protocol in the WAP architecture. The WDP layer operates above the data capable bearer services supported by the various network types. WDP offers a consistent service to the upper layer protocols of WAP and communicates transparently over one of the available bearer services.

The WDP provides a common interface to the upper layer protocols, and the security, session, and application layers can function independently of the underlying wireless network. The transport layer is adapted to specific features of the underlying bearer. The global interoperability can be achieved by using mediating gateways.

The WAP protocols operate over different bearer services, including short message, circuit switched data, and packet data. The bearers offer different quality of service with respect to throughput, error rate, and delays. The WAP protocols are designed to compensate for or tolerate these varying levels of service.

The WAP-layered architecture enables other services and applications to use the features of the WAP stack through a set of well-defined interfaces. External applications can access the session, transaction, security, and transport layers directly. This allows the WAP stack to be used for applications and services not currently specified by WAP, but valuable for the wireless market.

Figure 5.2 shows possible protocol stacks using WAP technology. The first stack represents a WAP application, a WAE user agent, running over WAP technology. The next stack

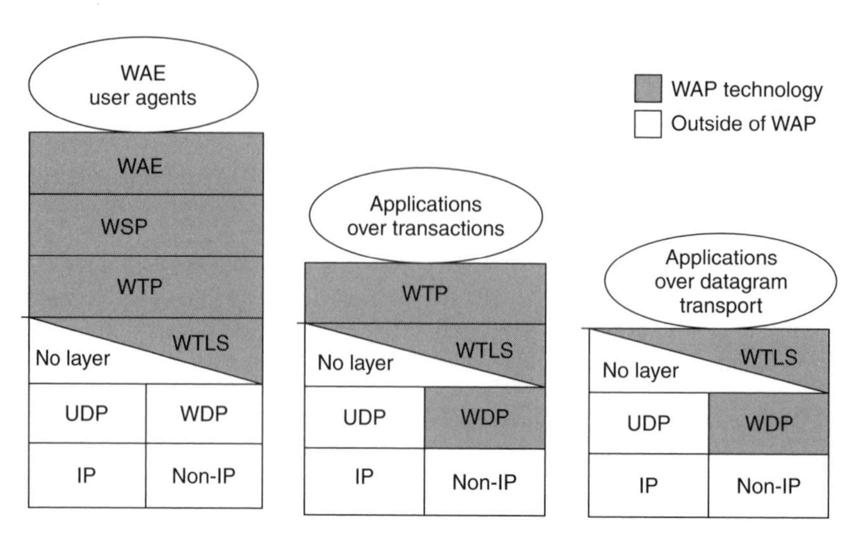

Figure 5.2 WAP stacks.

shows applications and services that require transaction services with or without security. The last stack shows applications and services that only require datagram transport with or without security.

5.6 SUMMARY

Mobile networks are growing in complexity and the cost of providing new value-added services to wireless users is increasing. To meet the requirements of mobile network operators, solutions must be: interoperable, scalable, efficient, reliable, and secure.

The WAP Forum is an industry group dedicated to the goal of enabling sophisticated telephony and information services on handheld wireless devices such as mobile telephones, pagers, PDAs and other WTs. Recognizing the value and utility of the World Wide Web architecture, the WAP Forum has chosen to align certain components of its technology very tightly with the Internet and the WWW. The WAP specifications extend and leverage mobile networking technologies (such as digital data networking standards) and Internet technologies, such as IP, HTTP, XML, URLs, scripting, and other content formats.

Structured data such as spreadsheets, address books, configuration parameters, financial transactions, technical drawings, and so on are often stored on a disk, for which they can use either a binary format or a text format. The latter allows the user, if necessary, to look at the data without the program that produced it. XML is a set of rules, guidelines, and conventions for designing text formats for such data, in a way that produces files that are easy to generate and read by a computer, that are unambiguous and that avoid common pitfalls such as lack of extensibility, lack of support for internationalization/localization, and platform-dependency.

The XP specification must define the concept of an envelope or outermost syntactical construct or structure within which all other syntactical elements of the message must be enclosed. The envelope must be described with XML Schema.

The WAP protocols operate over different bearer services, including short message, circuit switched data, and packet data. The bearers offer different quality of service with respect to throughput, error rate, and delays. The WAP protocols are designed to compensate for or tolerate these varying levels of service.

The WAP-layered architecture enables other services and applications to use the features of the WAP stack through a set of well-defined interfaces. External applications can access the session, transaction, security, and transport layers directly. This allows the WAP stack to be used for applications and services not currently specified by WAP, but valuable for the wireless market.

PROBLEMS TO CHAPTER 5

Protocols for wireless applications

Learning objectives

After completing this chapter, you are able to

- demonstrate an understanding of mobile and wireless devices;
- explain the features of mobile and wireless devices;
- explain how to access the Web;
- explain the protocols, and applications;
- demonstrate an understanding of WAP.
- explain the role of WAE;
- explain WSP, WTP, and WDP;
- explain WTLS.

Practice problems

5.1: What are the features of mobile and wireless devices?
5.2: What are the special considerations for mobile devices when it comes to using Web information?
5.3: Why are XML files text files?
5.4: What is WAP architecture?
5.5: What is the functionality of the WAE microbrowser environment?
5.6: What is the role of WSP?
5.7: What is the role of WTP?
5.8: What is the role of WTLS?
5.9: What is the role of WDP?

Practice problem solutions

5.1: Mobile and wireless devices are usually handheld devices, and accessing the World Wide Web presents a more constrained computing environment compared to desktop computers because of fundamental limitations of power and form factor. Mass-market handheld wireless devices tend to have: less powerful CPUs, less memory (both ROM and RAM), restricted power consumption, smaller displays, and different input devices (e.g., a phone keypad, voice input, etc.).

5.2: Mobile devices need special consideration when it comes to using Web information. Their displays are generally much smaller than a conventional computer screen and are capable of showing only a small amount of text. On a cellular phone, for example, there may be only enough space for three or four rows of text. Palmtop pocket-sized computers have screens smaller than a PC or laptop, but large enough to read e-mail (electronic mail) and documents with a small amount of text. Mobile devices have limited memory and processing speeds, and these considerations also need to be taken into account.

5.3: XML files are text files because that allows experts, such as programmers, to more easily debug applications, and in emergencies they can use a simple text editor to fix a broken XML file. But the rules for XML files are much stricter than for HTML. A forgotten tag, or an attribute without quotes makes the file unusable, while in HTML such practice is often explicitly allowed, or at least tolerated. In the official XML specification, the applications are not allowed to try to second-guess the creator of a broken XML file; if the file is broken, an application has to stop right there and issue an error message.

5.4: The WAP architecture provides a scalable and extensible environment for application development for mobile communication devices. The WAP protocol stack has a layered design, and each layer is accessible by the layers above, and by other services and applications. The WAP-layered architecture enables other services and applications to use the features of the WAP stack through a set of well-defined interfaces. External applications can access the session, transaction, security, and transport layers directly.

5.5: The WAE is a general-purpose application environment based on the combination of WWW and Mobile Telephony technologies. The WAE allows operators and service providers to build applications and services that can reach wireless platforms in an efficient and useful manner. WAE contains a microbrowser environment containing the following functionality:

- *Wireless Markup Language (WML)*: a lightweight markup language, similar to HTML, and optimized for use in handheld mobile devices;
- *WMLScript*: a lightweight scripting language, similar to JavaScript;
- *Wireless Telephony Application (WTA)*: telephony services and programming interfaces; and
- *Content formats*: a set of well-defined data formats, including images, phone book records, and calendar information.

5.6: WSP provides the application layer of WAP with an interface for two session services. The connection oriented service operates above the WTP. The connectionless service operates above a secure or nonsecure WDP.

5.7: The WTP runs on top of a datagram service and is suitable for implementation of MSs as a lightweight transaction oriented protocol. WTP operates efficiently over secure or nonsecure wireless datagram networks.

5.8: WTLS is a security protocol based upon the industry standard TLS protocol, formerly known as SSL. WTLS is used with the WAP transport protocols and is optimized for use over narrow-band communication channels.

WTLS can also be used for secure communication between terminals, for authentication of electronic business card exchange. The applications can selectively enable or disable WTLS features depending on their security requirements and the characteristics of the underlying network.

5.9: WDP is the transport layer protocol in the WAP architecture. The WDP layer operates above the data capable bearer services supported by the various network types. WDP offers a consistent service to the upper layer protocols of WAP and communicates transparently over one of the available bearer services.

6

Network architecture supporting wireless applications

The Wireless Application Environment (WAE) architecture is designed to support Mobile Terminals (MTs) and network applications using different languages and character sets. WAE user agents have a current language and accept content in a set of well-known character encoding sets. Origin server-side applications can emit content in one or more encoding sets and can accept input from the user agent in one or more encoding sets.

Wireless Telephony Application (WTA) is an application framework for telephony services. The WTA user agent has the capability for interfacing with mobile network services available to a mobile telephony device, that is, setting up and receiving phone calls.

The Wireless Application Protocol (WAP) Push framework introduces a means within the WAP effort to transmit information to a device without a previous user action. In the client/server model, a client requests a service or information from a server, which transmits information to the client. In this pull technology, the client pulls information from the server.

6.1 WAE ARCHITECTURE

The WAE architecture includes networking schemes, content formats, programming languages, and shared services. Interfaces are not standardized and are specific to a particular implementation. WAE can work with a browser and a class of user agents used in the World Wide Web (WWW).

In the Internet WWW, applications present content to a client in a set of standard data formats that are browsed by client side user agents known as *Web browsers*. A user agent sends requests for one or more data objects or content to an origin server, which responds with the requested data expressed in one of the standard formats known to the user agent [i.e., Hypertext Markup Language (HTML)]. The WWW logical model is shown in Figure 6.1.

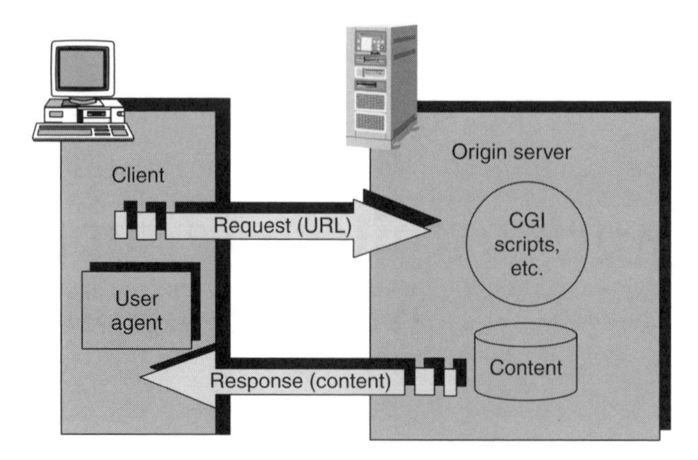

Figure 6.1 WWW logical model.

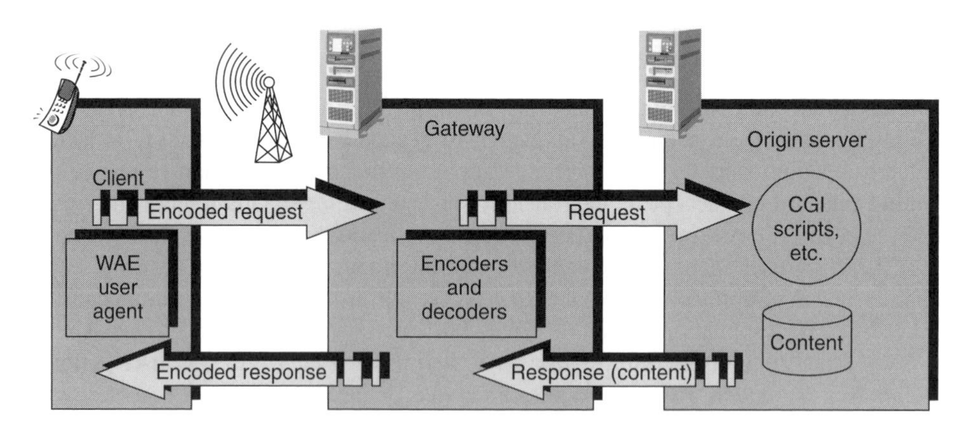

Figure 6.2 WAE logical model.

All resources on the WWW are named using Internet standard Uniform Resource Locators (URLs). All classes of data on the WWW are given as specific types, allowing the user agent to correctly distinguish and present them appropriately. The WWW defines a variety of standard content formats supported by most browser user agents, including the HTML, the JavaScript scripting language, and other formats like bitmap image formats. The WWW defines a set of standard networking protocols allowing any browser to communicate with any origin server, for example, Hypertext Transfer Protocol (HTTP).

The WAE logical model is shown in Figure 6.2. In the WAE model, the content is transported using standard protocols in the WWW domain and an optimized HTTP-like protocol in the wireless domain. The content and services in WAE architecture are hosted on standard Web origin servers using proven technologies like Common Gateway Interface (CGI). The content is located by using WWW standard URLs. WAE supports

Mobile Network Services such as Call Control and Messaging. WAE architecture supports low bandwidth and high latency networks and considers CPU processing constraints in MTs. WAE assumes the existence of gateway functionality responsible for encoding and decoding data transferred from and to the mobile client. The purpose of the encoding content delivered to the client is to minimize the size of data sent to the client Over The Air (OTA), and to minimize the computational energy required by the client to process the data. The gateway functionality can be added to origin servers or placed in dedicated gateways.

The main elements of the WAE model are WAE user agents, content generators, standard content encoding, and WTA. WAE user agents interpret network content referenced by a URL. Content generators are the applications or services on origin servers, like CGI scripts, that produce standard content formats in response to requests from user agents in MTs. Standard content encoding allows a WAE user agent to navigate Web content. WTA is a collection of telephony-specific extensions for call and feature control mechanisms providing advanced Mobile Network Services.

WAE is based on the architecture used for WWW proxy servers. The situation in which a user agent, a browser, must connect through a proxy to reach an origin server, the server that contains the desired content, is very similar to the case of a wireless device accessing a server through a gateway. Most connections between the browser and the gateway use WAP Session Protocol (WSP), regardless of the protocol of the destination server. URL refers only to the destination server's protocol and has no bearing on what protocols may be used in intervening connections. The gateway performs protocol conversion by translating requests from WSP into other protocols, and translating the responses back into WSP. Content conversion performed by the gateway is analogous to HTML/HTTP proxies available on the Web. In the HTTP scheme, the browser communicates with the gateway using WSP. The gateway provides protocol conversion functions to connect to an HTTP origin server.

WAE logical layers include user agents such as browsers, phone books, message editors, and so on, and services and formats including common elements and formats accessible to user agents such as Wireless Markup Language (WML), WMLScript, image formats, vCard (electronic business card) and vCalendar (electronic calendar and scheduling exchange) formats, and so on. The WAE client components are shown in Figure 6.3.

WAE allows the integration of domain-specific user agents with varying architectures and environments. A WTA user agent is specified as an extension to the WAE specification for the mobile telephony environments. The WTA extensions allow for accessing and interacting with mobile telephone features, like call control, and other applications assumed on the telephones, such as phone books and calendar applications. The features and capabilities of a user agent are decided by those who implement them.

WAE services and formats include the WML, the WMLScript (Wireless Markup Scripting Language), WAE applications, and WAE-supported content formats.

WML is a tag-based document language. It is an application of a generalized markup language and is specified as an Extensible Markup Language (XML) document type. WML is optimized for specifying presentation and user interaction on limited capability devices such as telephones and wireless MTs. WML and its supporting environment are designed using certain small narrow-band device constraints including small displays,

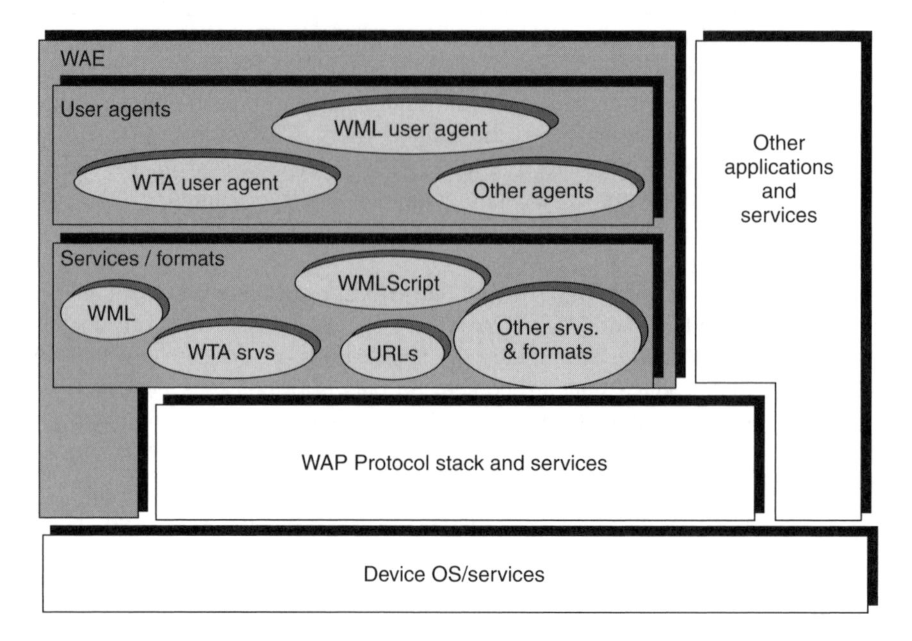

Figure 6.3 WAE client components.

limited user input facilities, narrow-band network connections, limited memory resources, and limited computational resources.

The WML features include

- support for text and images;
- support for user input;
- navigation and history stack;
- international support;
- Man–Machine Interface (MMI) independence;
- narrow-band optimization; and
- state and context management.

WMLScript is a lightweight procedural scripting language enhancing the standard browsing and presentation facilities of WML with behavioral capabilities, supporting more advanced user interface, adding intelligence to the client, providing a convenient mechanism to access the device and its peripherals, and reducing the need for round trips to the origin server. WMLScript is an extended subset of JavaScript for narrow-band devices and is integrated with WML for future services and in-device applications.

WAE user agents can use URL services. WAE components extend the URL semantics, for example, in WML, in which URL fragments are extended to allow linking to particular WMLScript functions. WAE allows formats for data types including images, multipart messages, and user agent-specific formats.

WML user agent logical architecture is shown in Figure 6.4. A user submits a request to the origin server using a WML user agent, which requests the service by using a URL

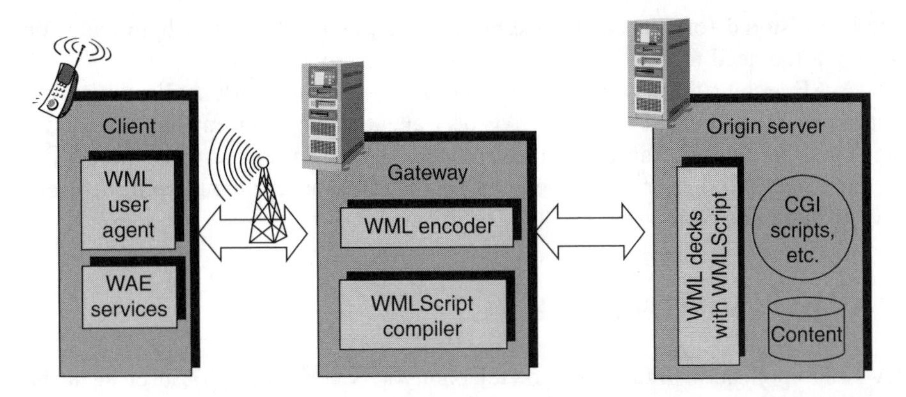

Figure 6.4 WML user agent logical architecture with gateway.

scheme operation. The origin server replies by sending a single deck in a textual format. On their way back to the client, textual decks are expected to pass through a gateway where they are converted into formats better suited for OTA transmission and limited-device processing. The gateway does all the necessary conversions between the textual and binary formats. A WML encoder (or tokenizer) in the gateway converts each WML deck into its binary format. Encoded content is then sent to the client to be displayed and interpreted.

The user agent may submit one or more additional requests, using a URL scheme, for WMLScript as the user agent encounters references to them in a WML deck. On its way back, a WMLScript compiler takes the script as input and compiles it into byte code that is designed for low bandwidth and thin mobile clients. The compiled byte code is then sent to the client for interpretation and execution.

Figure 6.5 shows WML user agent logical architecture without a gateway. WAE does not specify the location where the actual encoding and compilation is done. The origin servers may have built-in WML encoders and WMLScript compilers. Some services may

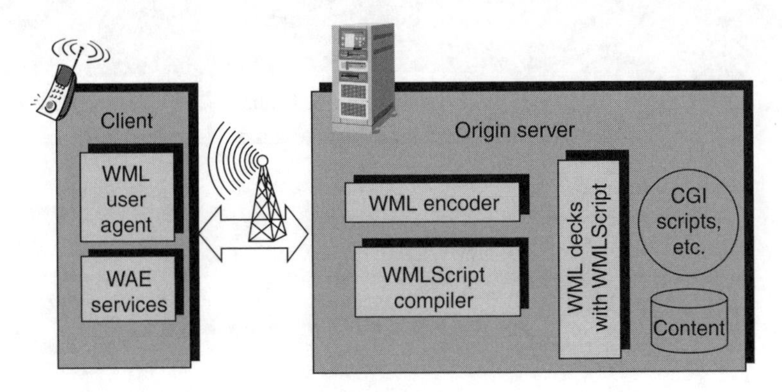

Figure 6.5 WML user agent logical architecture without a gateway.

be statically stored (or cached) in tokenized WML and WMLScript byte code formats, eliminating the need to perform fast conversion of the deck.

The WAE architecture is designed to support MTs and network applications using different languages and character sets. WAE user agents have a current language and accept content in a set of well-known character encoding sets. Origin server–side applications can emit content in one or more encoding sets and can accept input from the user agent in one or more encoding sets.

6.2 WTA ARCHITECTURE

WTA is an application framework for telephony services. The WTA user agent has the capability for interfacing with mobile network services available to a mobile telephony device, that is, setting up and receiving phone calls. Figure 6.6 shows a configuration of the WTA architecture. In this figure, the WTA user agent, the repository (persistent storage), and WTA Interface (WTAI) interact with each other and the other entities in a WTA-capable mobile client device. The WTA user agent is able to retrieve content from the repository and WTAI. This ensures that the WTA user agent can interact with mobile network functions like setting up calls and device-specific features like using the phonebook. The WTA user agent receives network events that can be bound to content, thus enabling dynamic telephony applications.

Network events available to the WTA user agent are the result of actions taken by services running in the WTA user agent itself. Telephony events initiated from outside the device are also passed to the WTA user agent and the network text message events

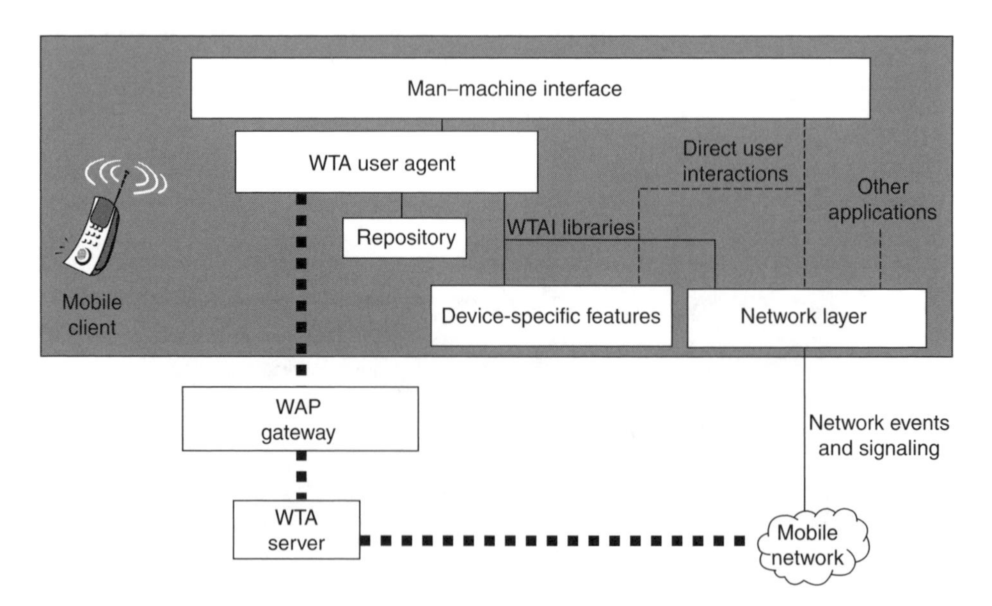

Figure 6.6 WTA architecture.

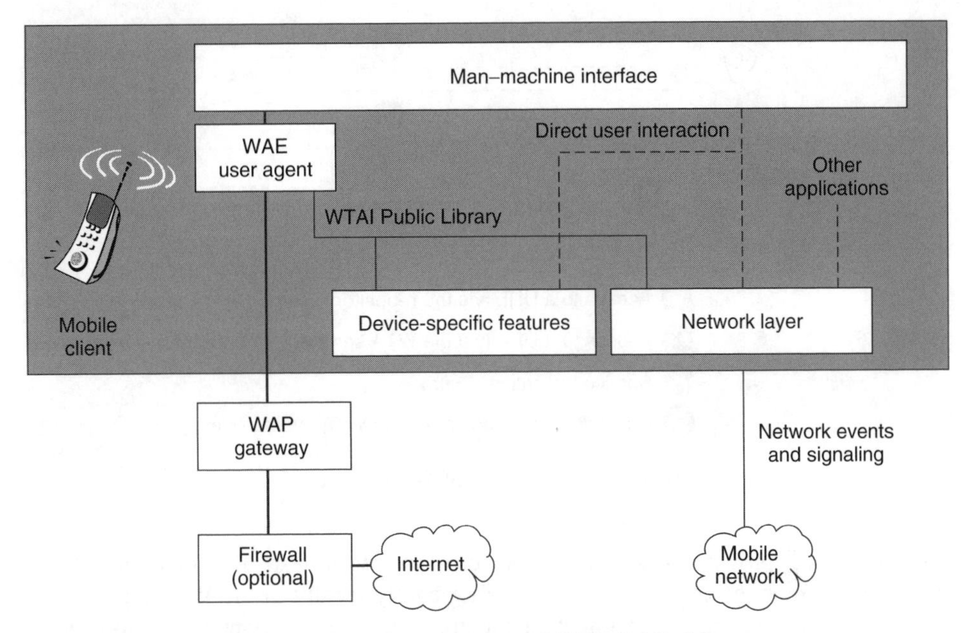

Figure 6.7 WAE user agent and WTA Public Library.

that are explicitly directed toward another user agent. The network events caused by the WML user agent do not affect the WTA user agent.

WTAI Public Library contains functions that can be called from any WAE application as shown in Figure 6.7 and provides access to telephone functionality. This library allows WML authors to include click-to-phone functionality within their content, to avoid users typing the number by using the default MMI.

In Figure 6.7, the WAE user agent and WTAI Public Library interact with each other and the other entities in a WTA-capable mobile client. The WAE user agent only retrieves its content *via* the WAP gateway and only has access to the WTAI Public Library functions. These functions expose simple functionality such as the ability to place a call, but do not allow fully featured telephony control. Only a WTA user agent is able to fully control the telephony features of the device. The WAE user agent is not able to receive and react to telephony and network text events.

Figures 6.6 and 6.7 show logical separations of the two user agents. They can coexist on the same device and are likely to be implemented with common code elements.

The WTA server is a Web server delivering content requested by a client. A WTA user agent, like an Internet Web browser, uses URLs to reference content on the WTA server. A URL can be used to reference an application on a Web server, for instance, a CGI script, that is executed when it is referenced. The applications can be programmed to perform a wide range of tasks, for example, generate dynamic content and interact with external entities. By referencing applications on a WTA server, it is possible to create services that use URLs to interact with the mobile network, such as an Intelligent Network node, and other entities, such as a voice mail system. The concept of referencing applications

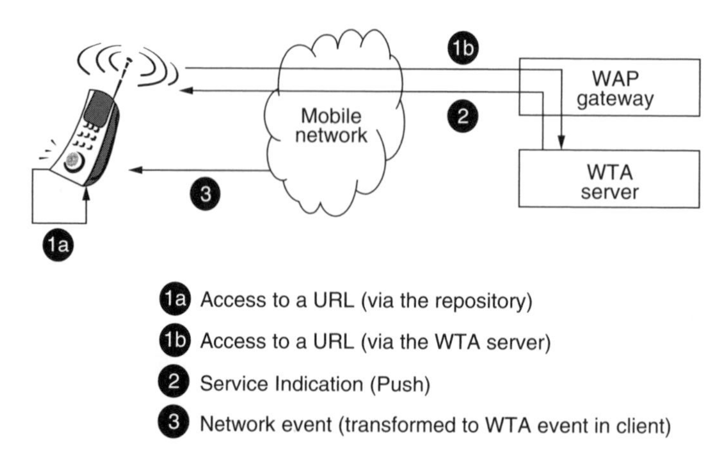

Figure 6.8 Initiation of WTA services.

on a WTA server provides a simple and powerful model on how to seamlessly integrate services in the mobile network with services executing locally in the WAP client.

WTA services appear to the client in the form of various content formats, such as WTA–WML, WMLScript, and so on. The WTA user agent executes content that is persistently stored in the client's repository or content retrieved from a WTA server. The WTA user agent can act on events from the mobile network, for instance, an incoming call.

Figure 6.8 shows how to initiate a WTA service in the WTA user agent. The WTA user agent executes content within the boundary of a well-known context. The service defines the extent of a context and its associated content. The start of a service is marked by the initiation of a new context, and the termination of a context marks the end of a service.

The repository is a persistent storage module within the MT that may be used to eliminate the need for network access when loading and executing frequently used WTA services. The repository also addresses the issue of how a WTA service developer ensures that time-critical WTA events are handled in a timely manner. The repository addresses the issues of how the WTA services developer preprogram the device with content, and how the WTA services developer improves the response time for a WTA service.

The repository can be accessed by a service using one of the following methods:

- A WTA event associated with a channel is detected, and the user agent invokes a URL as specified by the associated channel;
- The end user accesses services stored in the repository through an implementation-dependent representation (for instance, a menu containing the labels of the channels) of the allowed services (channels explicitly specified as user accessible by the channel definition) in the repository;
- The content of URL retrieved from the repository may be given to the user agent by providing the URL in content or delivering it by Service Indication (SI).

The WTA applications, that is, content loaded or otherwise received from the WTA server, may access the repository.

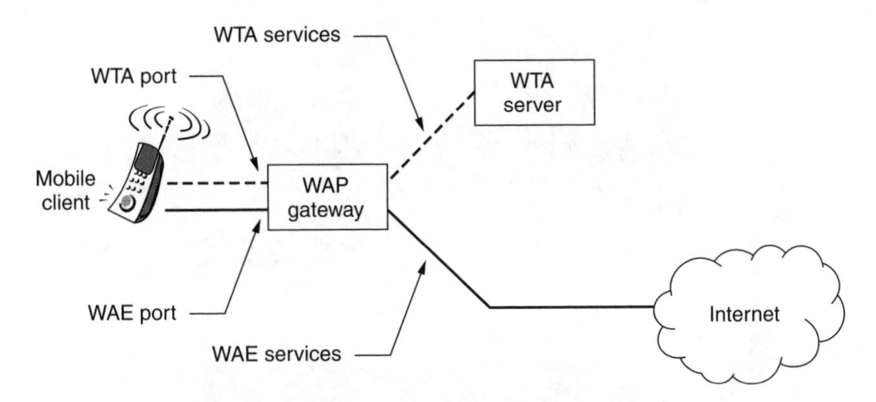

Figure 6.9 WDP port numbers and access control.

A WTA service invoking WTAI functions enables access to local functions in the mobile client. These functions allowing for setting up calls, and accessing the users local phone book, must ensure that only authorized WTA services are permitted to execute. The trusted mobile telephony service provider, which provides an acceptable level of security in the network, can choose to run all WTA services itself not allowing other providers or it can choose to delegate the administration of its WTA services to a third party. The Wireless Datagram Protocol (WDP) uses predefined port numbers to separate a WTA service from a common WAE service as shown in Figure 6.9. A WTA session established by the WTA user agent must use one of the dedicated, secure WTA ports on the gateway. The WTA user agent must not retrieve WTA content outside the WTA session. WTA content received outside the WTA session and Service Indication addressing the WTA user agent but delivered outside a WTA session shall be discarded.

The repository is used to store WTA content persistently. This provides a mechanism that ensures timely handling of content related to WTA services initiated by WTA events and has the following characteristics:

- The repository contains a set of channels and resources.
- Resources are data downloaded with WSP (that is, WTA–WML deck) and are stored along with their metadata, that is, content type and the HTTP 1.1 entity tag, and location (URL).
- A channel is a resource that contains a set of links and resources and has identity and freshness.
- Channels in the repository have a freshness lifetime (the HTTP 1.1 expiry date header), beyond which time they are considered stale. Stale channels are subject to automatic removal by the user agent. Resources are subject to automatic removal from the repository if the channel does not reference them.
- If the repository contains a channel that is not stale, it is guaranteed that the repository contains all resources named in that channel. The loading and unloading of a channel is an atomic operation in that no user agent will recognize the presence of the channel until all the content in the channel has been successfully stored in the repository.

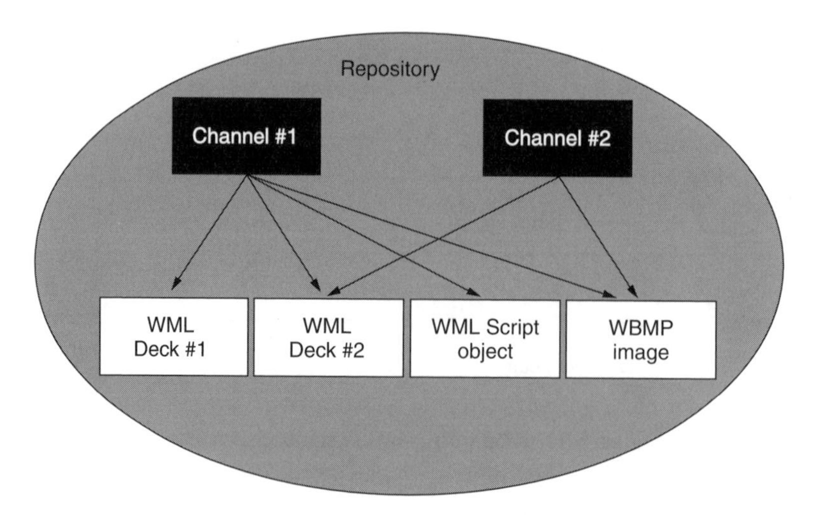

Figure 6.10 Repository.

- A label may be associated with a channel to give a textual description of the service indicated by the channel.

Resources in the repository may be referenced by more than one channel. A resource is present in the repository if one or more channels reference it. Figure 6.10 shows how channels may share resources stored in the repository.

WTA services are created using WTA–WML and WMLScript. Telephony functions can be accessed from WMLScript through the WTAI, which also provides access to telephony functions from WTA–WML by using Uniform Resource Identifier (URIs). URIs form a unifying naming model to identify features independently of the internal structure of the device and the mobile network. The WTA services reside on the WTA server. The client addresses WTA services by using URLs.

Examples of WTA services include

- *Extended set of user options for handling incoming calls (incoming call section)*: The service is started when an incoming call is detected in the client. A menu with user options is presented to the user. Examples of options are

  Accept call
  Redirect to voice mail
  Redirect to another subscriber
  Send special message to caller.

- *Voice mail*: The user is notified that he or she has voice mails and retrieves a list of them from the server. The list is presented on the client's display. When a certain voice mail has been selected, the server sets up a call to the client and the user listens to the selected voice mail.

- *Call subscriber from message list or log*: When a list of voice, fax, or e-mails or any kind of call log is displayed, the user has the option of calling the originator of a selected entry in the list or log.

The incoming call selection service is started when an incoming call is detected in the client and a menu with various call-handling options is presented to the user as shown in Figure 6.11. A valid channel and its associated content are stored in the repository. The client is not engaged in any other WTA service (i.e., no temporary event bindings exist).

The following events in this example of incoming call selection are shown in Figure 6.11.

1. The mobile network receives an incoming call and sends a Call Indication to the mobile subscriber.
2. In the client, the incoming call WTA event (wtaev-cc/ic) is generated. The repository is checked to find a dedicated channel. The channel provides the URL to the Incoming Call Selection service stored in the repository.
3. The user agent requests the content from the repository.
4. The repository returns the requested content.
5. The content is loaded into a clean context and starts executing. The service presents to the user a list of options, from which he or she can choose how to proceed with a call in progress. In this example, the user elects to answer the call. The WTAI function WTAIVoiceCall.accept is invoked.

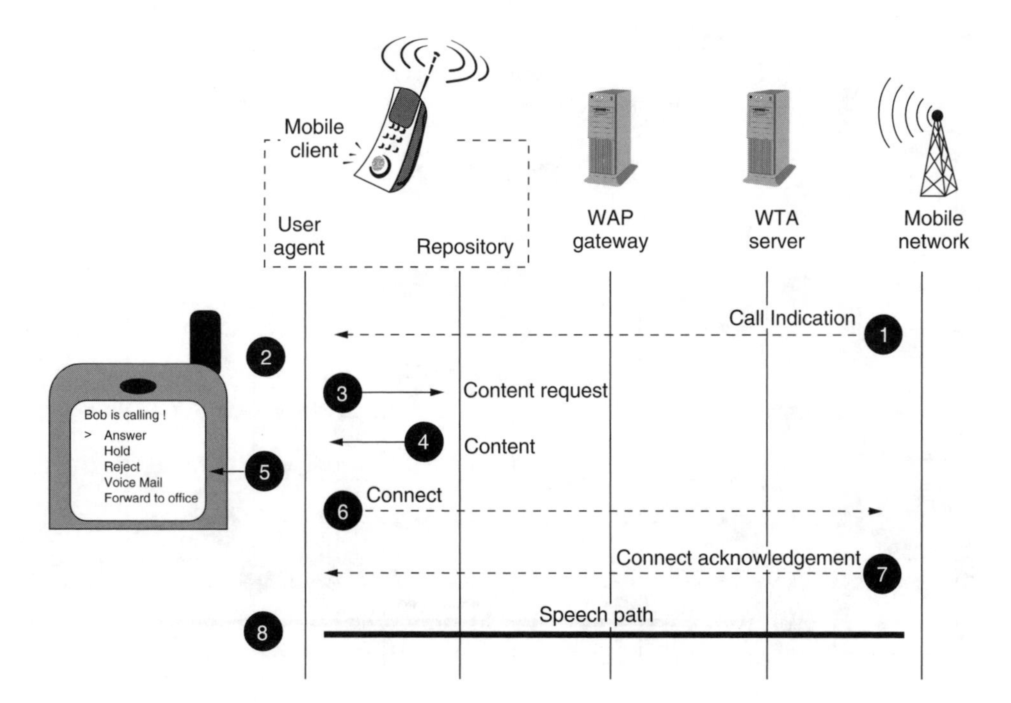

Figure 6.11 Incoming call selection.

6. A Connect request is sent to the mobile network (the invoked WTAI function communicates with the mobile network).
7. A Connect Acknowledgement (ACK) is generated in the mobile network. A result code indicating the outcome of the calls is generated internally in the phone.
8. A speech path between the mobile network and the client is established.

Figure 6.12 shows how the voice mail service is established within the WTA framework. The user is notified that he or she has received new voice mails, and the user chooses to listen to one of them.

The following events in this example of voice mail are shown in Figure 6.12.

1. The Voice Mail System notifies the WTA server that there are new voice mails. A list of voice mails is also sent to the WTA server.
2. The WTA server creates new service content on the basis of the list received from the voice mail system. The content is stored on the server and its URL is included in an SI that will be pushed to the client. The SI's message reads: 'You have 3 new voice mails'.
3. The WAP gateway sends the SI to the client using push.

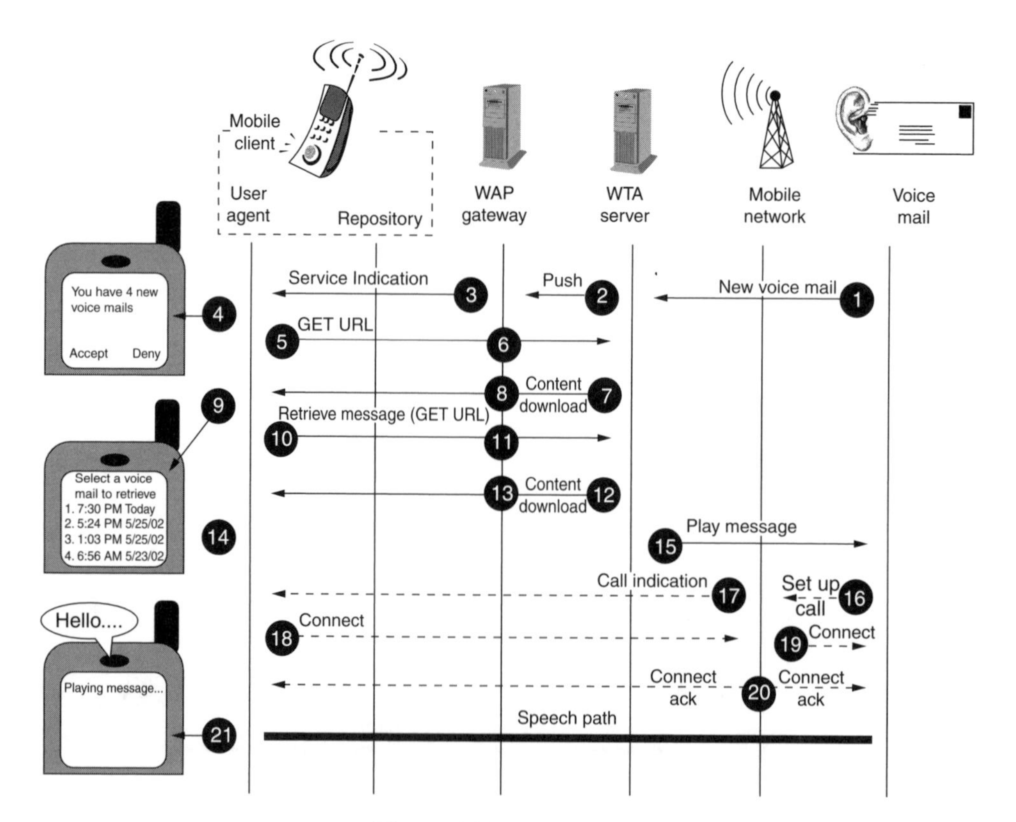

Figure 6.12 Voice mail.

4. The user is notified about the SI by a message delivered with the SI. The user chooses to accept the SI.
5. A WSP Get request is sent to the WAP gateway (URL provided by the SI).
6. The WAP gateway makes a WSP/HTTP conversion.
7. The WTA server returns the earlier created voice mail service.
8. The WAP gateway makes an HTTP/WSP conversion.
9. The voice mail service is now executing in the client. The user is presented with a list of voice mails originating from the Voice Mail System (a WTA–WML Select List is created in Step 2). The user selects a certain voice mail to listen to.
10. Another WSP Get request is sent to the WAP gateway. The requested deck identifies the selected voice mail.
11. The WAP gateway makes a WSP/HTTP conversion.
12. The WTA server returns the requested deck. The deck only contains one card with a single WTA–WML task. The URL is automatically called when the card is executed and it refers to a card in the earlier downloaded voice mail content that binds the incoming call event (wtaev-cc.ic) so that the subsequent call from the Voice Mail System will be answered automatically. Now, the WTA server is also informed about which voice mail the user has chosen to retrieve.
13. The WAP gateway makes an HTTP/WSP conversion.
14. The incoming call event (wtaev-cc/ic) is temporarily bound so that the call from the Voice Mail System will be answered automatically. To avoid that the voice mail service answers a call from someone other than the voice mail system, the calling party's phone number (callerId parameter of the wtaev-cc/ic event) should be preferably checked.
15. The WTA server instructs the Voice Mail System to play the selected voice mail.
16. The Voice Mail System instructs the mobile network to set up a call to the client.
17. The mobile network sets up a call to the client.
18. The client answers the call automatically as a result of the content loaded in Steps 12 to 14.
19. The mobile network informs the Voice Mail System that the client has accepted the call.
20. ACKs are sent to the client and the Voice Mail System.
21. A speech path is established between the Voice Mail System and the client, and the message is played.

6.3 WAP PUSH ARCHITECTURE

The WAP Push framework introduces a means within the WAP effort to transmit information to a device without a previous user action. In the client/server model, a client requests a service or information from a server, which transmits information to the client. In this pull technology, the client pulls information from the server. An example of pull technology is WWW, in which a user enters a URL (the request), sent then to a server, which answers by sending a Web page (the response) to the user. In the push technology based on client/server model, there is no explicit request from the client before the server

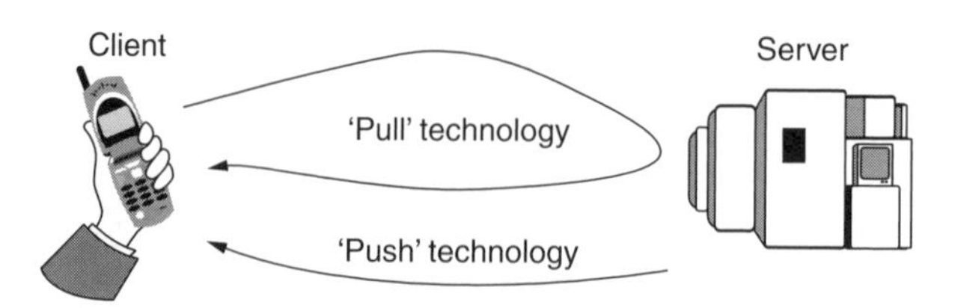

Figure 6.13 Comparison of pull and push technology.

transmits its content. Figure 6.13 illustrates pull and push technology. Pull transactions are initiated by the client, whereas push transactions are initiated by the server.

A push operation in WAP occurs when a Push Initiator transmits content to a client using either the Push OTA protocol or the Push Access Protocol (PAP). The Push Initiator does not share a protocol with the WAP client since the Push Initiator is on the Internet and the WAP client is on the WAP domain. The Push Initiator contacts the WAP Client through a translating Push Proxy Gateway (PPG) from the Internet side, delivering content for the destination client using Internet protocols. The PPG forwards the pushed content to the WAP domain, and the content is then transmitted over the air in the mobile network to the destination client. The PPG may be capable of notifying the Push Initiator about the final outcome of the push operation, and it may wait for the client to accept or reject the content in two-way mobile networks. It may also provide the Push Initiator with client capability lookup services by letting a Push Initiator select the optimal content for this client.

The Internet side PPG access protocol is called *the PAP*. The WAP side protocol is called *OTAProtocol*. The PAP uses XML messages that may be tunneled through various Internet protocols, for example, HTTP. The OTA protocol is based on WSP services. The Push framework with the protocols is shown in Figure 6.14.

The PPG acts as an access point for content pushes from the Internet to the mobile network, and associated authentication, security, client control, and so on. The PPG owner decides the policies about who is able to gain access to the WAP network, who is able to push content, and so on. The PPG functionality may be built into the pull WAP gateway that gives the benefit of shared resources and shared sessions over the air.

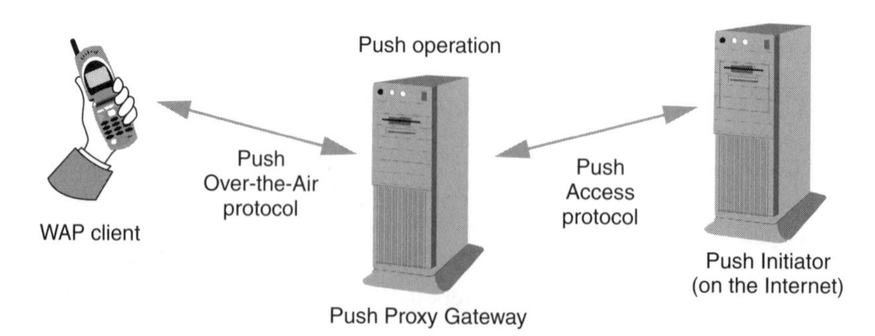

Figure 6.14 Push framework with protocols.

The PPG performs the following services:

- Push Initiator and authentication; access control;
- parsing of and error detection in content control information;
- client discovery services;
- address resolution;
- binary encoding and compilation of certain content types to improve efficiency OTA;
- protocol conversion.

The PPG accepts pushed content from the Internet using the PAP. The PPG acknowledges successful parsing or reports unsuccessful parsing of the control information and may report debug information about the content. It may also perform a callback to the pushing server when the final status of the push submission has been reached, if the Push Initiator so requests.

When the content has been accepted for delivery, the PPG attempts to find the correct destination device and deliver the content to the client using the Push OTA protocol. The PPG attempts to deliver the content until a timeout expires, which can be set by the Push Initiator and/or the policies of the mobile operator.

The PPG may encode WAP content types into their binary counterparts. This transaction takes place before delivery over the air. Other content types may be forwarded as received. The Push Initiator may also precompile its content into binary form to take workload off the PPG, for example. When the PPG receives precompiled WML, WMLScript, or SIs, they are forwarded as received.

The PPG may implement addressing aliasing schemes to enable special multi- and broadcast cases, in which special addresses may translate to a broadcast operation.

A Push Initiator may query the PPG for client capabilities and preferences to create better formatted content for a particular WAP device.

The PAP is used by an Internet-based Push Initiator to push content to a mobile network addressing its PPG. The PAP initially uses HTTP, but it can be tunneled through any other or future Internet protocol. The PAP carries an XML-style entity that may be used with other components in a multipart-related document.

The PAP supports the following operations:

- Push Submission (Initiator to PPG)
- Result Notification (PPG to Initiator)
- Push Cancellation (Initiator to PPG)
- Status Query (Initiator to PPG)
- Client Capabilities Query (Initiator to PPG).

The push message contains three entities: a control entity, a content entity, and optionally a capability entity. They are used in a multipart-related message, which is sent from the Push Initiator to the PPG. The control entity is an XML document containing delivery instructions destined for the PPG, and the content entity is destined for the mobile device.

If the Push Initiator requested a confirmation of successful delivery, the message is transmitted from the PPG to the Push Initiator when the content is delivered to the mobile device over a two-way bearer, or transmitted to the device over a one-way bearer, and it contains an XML entity. The message is also transmitted in case of a detected delivery failure to inform the Initiator about it.

The Push Initiator relies on the response from the PPG; a confirmed push is then confirmed by the WAP device only when the target application has taken responsibility for the pushed content. Otherwise, the application must abort the operation and the Push Initiator knows that the content never reached its destination.

An XML entity can be transmitted from the Push Initiator to the PPG requesting cancellation of the previously submitted content. The PPG responds with an XML entity whether or not the cancellation was successful. An XML can also be transmitted from the Push Initiator to the PPG requesting status of the previously submitted content. The PPG responds with an XML entity. An XML entity transmitted from the Push Initiator to the PPG can request the capabilities of a device on the network. The PPG responds with a multipart related in two parts, in which the multipart root is the result of the request, and the second part is the capabilities of the device. The WAP is carried over HTTP/1.1 in this issue of WAP Push.

The SI content type provides the ability to send notifications to end users in an asynchronous manner. An SI contains a short message and a URI indicating a service. The message is presented to the end user upon reception, and the user is given the choice to either start the service indicated by the URI immediately or to postpone the SI for later handling. If the SI is postponed, the client stores it and the end user is given the possibility to act upon it at a later time.

The Push OTA protocol is a thin protocol layer on top of WSP, and it is responsible for transporting content from the PPG to the client and its user agents. The OTA protocol may use WSP sessions to deliver its content. Connection-oriented pushes require that an active WSP session is available, but a session cannot be created by the server. When there is no active WSP session, the Push framework introduces a Session Initiation Application (SIA) in the client that listens to session requests from the OTA servers and responds by setting up a WSP session for push purposes. The client may verify the identity information in this request against a list of recognized OTA servers before attempting to establish any push sessions. Push delivery may also be performed without the use of sessions in a connectionless manner, which is needed in one-way networks.

A connection-oriented push requires an active WSP session. Only the client can create sessions. If the server receives a request for a connection-oriented push to a client, and there are no active sessions to that client, the server cannot deliver the push content. A session request is sent to a special application in the client known as the SIA. This request contains information necessary for a client to create a push session. The SIA in the client after receiving a session request establishes a session with the PPG and indicates which applications accept content over the newly opened session. The SIA may also ignore the request if there is no suitable installed application as requested in the session request.

When a client receives pushed content, a dispatcher looks at the push message header to determine its destination application. This dispatcher is responsible for rejecting content that does not have a suitable destination application installed, and for confirming push operations to the PPG when the appropriate application takes responsibility for pushed content.

6.4 SUMMARY

The main elements of the WAE model are WAE user agents, content generators, standard content encoding, and wireless telephony applications. WAE user agents interpret network content referenced by a URL. Content generators are the applications or services on origin servers, like CGI scripts, that produce standard content formats in response to requests from user agents in MTs. Standard content encoding allows a WAE user agent to navigate Web content. WTA is a collection of telephony-specific extensions for call and feature control mechanisms providing advanced Mobile Network Services.

The repository is a persistent storage module within the MT that may be used to eliminate the need for network access when loading and executing frequently used WTA services. The repository also addresses the issue of how a WTA service developer ensures that time critical WTA events are handled in a timely manner. The repository addresses the issues of how the WTA services developer preprogram the device with content and how the WTA services developer improves the response time for a WTA service.

A push operation in WAP occurs when a Push Initiator transmits content to a client using either the Push OTA protocol or the PAP. The Push Initiator does not share a protocol with the WAP client since the Push Initiator is on the Internet and the WAP client is on the WAP domain. The Push Initiator contacts the WAP Client through a translating PPG from the Internet side, delivering content for the destination client using Internet protocols. The PPG forwards the pushed content to the WAP domain, and the content is then transmitted over the air in the mobile network to the destination client. The PPG may be capable of notifying the Push Initiator about the final outcome of the push operation, and it may wait for the client to accept or reject the content in two-way mobile networks. It may also provide the Push Initiator with client capability lookup services, letting a Push Initiator select the optimal content for this client.

The Internet side PPG access protocol is called the *PAP*. The WAP side protocol is called the *OTAProtocol*. The PAP uses XML messages that may be tunneled through various Internet protocols, for example, HTTP. The OTA protocol is based on WSP services.

PROBLEMS TO CHAPTER 6

Network architecture supporting wireless applications

Learning objectives

After completing this chapter, you are able to

- demonstrate an understanding of the network architecture supporting wireless applications;
- explain the role of WAE architecture;
- explain WTA architecture;
- explain WAP Push architecture.

Practice problems

6.1: What are the main elements of the WAE model?
6.2: What is the role of the repository in the WTA services?
6.3: What is the WAP Push framework?

Practice problem solutions

6.1: The main elements of the WAE model are WAE user agents, content generators, standard content encoding, and WTA. WAE user agents interpret network content referenced by a URL. Content generators are the applications or services on origin servers, like CGI scripts, that produce standard content formats in response to requests from user agents in MTs. Standard content encoding allows a WAE user agent to navigate Web content. WTA is a collection of telephony specific extensions for call and feature control mechanisms providing advanced Mobile Network Services.

6.2: The repository is a persistent storage module within the MT that may be used to eliminate the need for network access when loading and executing frequently used WTA services. The repository also addresses the issue of how a WTA service developer ensures that time-critical WTA events are handled in a timely manner. The repository addresses the issues of how the WTA services developer preprogram the device with content, and how the WTA services developer improves the response time for a WTA service.

6.3: The WAP Push framework introduces a means within the WAP effort to transmit information to a device without a previous user action. In the client/server model, a client requests a service or information from a server, which transmits information to the client. In this pull technology, the client pulls information from the server. An example of pull technology is WWW, in which a user enters a URL (the request) sent then to a server, which answers by sending a Web page (the response) to the user. In the push technology based on the client/server model, there is no explicit request from the client before the server transmits its content. Pull transactions are initiated by the client, whereas push transactions are initiated by the server. A push operation in WAP occurs when a Push Initiator transmits content to a client using either the Push OTA protocol or the PAP. The Push Initiator does not share a protocol with the WAP client since the Push Initiator is on the Internet and the WAP client is on the WAP domain. The Push Initiator contacts the WAP Client through a translating PPG from the Internet side, delivering content for the destination client using Internet protocols.

7

XML, RDF, and CC/PP

Extensible Markup Language (XML) describes a class of data objects called XML documents and partially describes the behavior of the computer programs that process them. XML is an application profile or restricted form of the Standard Generalized Markup Language (SGML).

Resource Description Framework (RDF) can be used to create a general, yet extensible framework for describing user preferences and device capabilities. This information can be provided by the user to servers and content providers. The servers can use this information describing the user's preferences to customize the service or content provided. The ability of RDF to reference profile information *via* URLs assists in minimizing the number of network transactions required to adapt content to a device, while the framework fits well into the current and future protocols.

A Composite Capability/Preference Profile (CC/PP) is a collection of the capabilities and preferences associated with user and the agents used by the user to access the World Wide Web. These user agents include the hardware platform, system software, and applications used by the user. User agent capabilities and references can be thought of as metadata or properties and descriptions of the user agent hardware and software.

7.1 XML DOCUMENT

XML documents are made up of storage units called entities, which contain either parsed or unparsed data. Parsed data is made up of characters, some of which form character data and some of which form markup. Markup encodes a description of the document's storage layout and logical structure. XML provides a mechanism to impose constraints on the storage layout and logical structure.

A software module called an XML processor is used to read XML documents and provide access to their content and structure. It is assumed that an XML processor is doing its work on behalf of another module called the application. An XML processor reads XML data and provides the information to the application.

The design goals for XML are

- to be straightforwardly usable over the Internet,
- to support a wide variety of applications,
- to be compatible with SGML,
- to create easy-to-write programs that process XML documents,
- to keep the number of optional features in XML to the absolute minimum, ideally zero,
- to have XML documents human-legible and reasonably clear,
- to prepare XML design quickly,
- to have the design of XML formal and concise,
- to have XML documents that are easy to create,
- to have terseness in XML markup of minimal importance.

A data object is an XML document if it is well formed, which may be valid if it meets certain further constraints. Each XML document has both a logical and a physical structure. Physically, the document is composed of units called entities. An entity may refer to other entities to cause their inclusion in the document. A document begins in a root or document entity. Logically, the document is composed of declarations, elements, comments, character references, and Processing Instructions (PIs), all of which are indicated in the document by explicit markup. The logical and physical structures must nest properly.

Matching the document production implies that it contains one or more elements, and there is exactly one element, called the root or document element, no part of which appears in the content of any other element. For all other elements, if the start-tag is in the content of another element, the end-tag is in the content of the same element. The elements, delimited by start- and end-tags, nest properly within each other.

A parsed entity contains text, a sequence of characters, which may represent markup or character data. Characters are classified for convenience as letters, digits, or other characters. A letter consists of an alphabetic or syllabic base character or an ideographic character. A Name is a token beginning with a letter or one of a few punctuation characters, and continuing with letters, digits, hyphens, underscores, colons, or full stops, together known as name characters. The Name spaces assign a meaning to names containing colon characters. Therefore, authors should not use the colon in XML names except for name space purposes, but XML processors must accept the colon as a name character. An Nmtoken (name token) is any mixture of name characters.

Literal data is any quoted string not containing the quotation mark used as a delimiter for that string. Literals are used for specifying the content of internal entities (EntityValue), the values of attributes (AttValue), and external identifiers (SystemLiteral). Note that a SystemLiteral can be parsed without scanning for markup.

Text consists of intermingled character data and markup. Markup takes the form of start-tags, end-tags, empty-element tags, entity references, character references, comments, Character Data (CDATA) section delimiters, document type declarations, processing instructions, XML declarations, text declarations, and any white space that is at the top level of the document entity (that is, outside the document element and not inside any other markup). All text that is not markup constitutes the character data of the document.

Comments may appear anywhere in a document outside other markup; in addition, they may appear within the document type declaration at places allowed by the grammar. They are not part of the document's character data; an XML processor may, but need not, make it possible for an application to retrieve the text of comments. For compatibility, the string " " (double-hyphen) must not occur within comments. Parameter entity references are not recognized within comments.

PIs allow documents to contain instructions for applications. PIs are not part of the document's character data, but must be passed through to the application. The PI begins with a target (PITarget) used to identify the application to which the instruction is directed. The target names XML, xml, and so on are reserved for specification standardization. The XML Notation mechanism may be used for formal declaration of PI targets. Parameter entity references are not recognized within PIs.

Markup declarations can affect the content of the document, as passed from an XML processor to an application; examples are attribute defaults and entity declarations. The stand-alone document declaration, which may appear as a component of the XML declaration, signals whether there are such declarations, which appear external to the document entity or in parameter entities. An external markup declaration is defined as a markup declaration occurring in the external subset or in a parameter entity (external or internal, the latter being included because nonvalidating processors are not required to read them).

In a stand-alone document declaration, the value 'yes' indicates that there are no external markup declarations that affect the information passed from the XML processor to the application. The value 'no' indicates that there are or may be such external markup declarations. The stand-alone document declaration only denotes the presence of external declarations; the presence, in a document, of references to external entities, when those entities are internally declared, does not change its stand-alone status. If there are no external markup declarations, the stand-alone document declaration has no meaning. If there are external markup declarations but there is no stand-alone document declaration, the value no is assumed.

Each XML document contains one or more elements, the boundaries of which are either delimited by start-tags and end-tags, or, for empty elements, by an empty-element tag. Each element has a type, identified by name, sometimes called its Generic Identifier (GI), and may have a set of attribute specifications. Each attribute specification has a name and a value.

An element is valid if there is a declaration matching element declaration in which the Name matches the element type, and one of the following holds:

1. The declaration matches EMPTY and the element has no content.
2. The declaration matches CHILDREN and the sequence of child elements belongs to the language generated by the regular expression in the content model, with optional white space between the start-tag and the first child element, between child elements, or between the last child element and the end-tag.
3. The declaration matches MIXED and the content consists of character data and child elements whose types match names in the content model.
4. The declaration matches ANY, and the types of any child elements have been declared.

The element structure of an XML document may, for validation purposes, be constrained using element-type and attribute-list declarations. An element-type declaration constrains the element's content. Element-type declarations often constrain which element types can appear as children of the element. At the user option, an XML processor may issue a warning when a declaration mentions an element type for which no declaration is provided, but this is not an error.

An element type has element content when elements of that type must contain only child elements (no character data), optionally separated by white space. In this case, the constraint includes a content model, a simple grammar governing the allowed types of the child elements and the order in which they are allowed to appear. The grammar is built on content particles, which consist of names, choice lists of content particles, or sequence lists of content particles.

Attribute-list declarations may be used

- to define the set of attributes pertaining to a given element type;
- to establish type constraints for these attributes;
- to provide default values for attributes.

Attribute-list declarations specify the name, data type, and default value (if any) of each attribute associated with a given element type.

An XML document may consist of one or many storage units. These are called *entities*; they all have content and are all [except for the document entity and the external Document Type Definition (DTD) subset] identified by entity name. Each XML document has one entity called the document entity, which serves as the starting point for the XML processor and may contain the whole document.

Entities may be either parsed or unparsed. A parsed entity's contents are referred to as its replacement text; this text is considered an integral part of the document. An unparsed entity is a resource whose contents may or may not be text, and if text, may be other than XML. Each unparsed entity has an associated notation, identified by name. Beyond a requirement that an XML processor makes the identifiers for the entity and notation available to the application, XML places no constraints on the contents of unparsed entities. Parsed entities are invoked by name using entity references – unparsed entities by name, given in the value of ENTITY or ENTITIES attributes.

General entities are entities for use within the document content. General entities are sometimes referred to with the unqualified term entity when this leads to no ambiguity. Parameter entities are parsed entities for use within the DTD. These two types of entities use different forms of reference and are recognized in different contexts. Furthermore, they occupy different name spaces; a parameter entity and a general entity with the same name are two distinct entities.

7.2 RESOURCE DESCRIPTION FRAMEWORK (RDF)

The RDF is a foundation for processing metadata; it provides interoperability between applications that exchange machine-understandable information on the Web. RDF uses

XML to exchange descriptions of Web resources but the resources being described can be of any type, including XML and non-XML resources. RDF emphasizes facilities to enable automated processing of Web resources. RDF can be used in a variety of application areas, for example, in resource discovery to provide better search engine capabilities; in cataloging for describing the content and content relationships available at a particular Web site, page, or digital library, by intelligent software agents to facilitate knowledge sharing and exchange; in content rating; in describing collections of pages that represent a single logical document; in describing intellectual property rights of Web pages; and in expressing the privacy preferences of a user as well as the privacy policies of a Web site. RDF with digital signatures is the key to building the Web of Trust for electronic commerce, collaboration, and other applications.

Descriptions used by these applications can be modeled as relationships among Web resources. The RDF data model defines a simple model for describing interrelationships among resources in terms of named properties and values. RDF properties may be thought of as attributes of resources and in this sense correspond to traditional attribute-value pairs. RDF properties also represent relationships between resources. As such, the RDF data model can therefore resemble an entity-relationship diagram. The RDF data model, however, provides no mechanisms for declaring these properties, nor does it provide any mechanisms for defining the relationships between these properties and other resources. That is the role of RDF Schema.

To describe bibliographic resources, for example, descriptive attributes including author, title, and subject are common. For digital certification, attributes such as checksum and authorization are often required. The declaration of these properties (attributes) and their corresponding semantics are defined in the context of RDF as an RDF schema. A schema defines not only the properties of the resource (e.g., title, author, subject, size, color, etc.) but may also define the kinds of resources being described (books, Web pages, people, companies, etc.).

The type system is specified in terms of the basic RDF data model – as resources and properties. Thus, the resources constituting this system become part of the RDF model of any description that uses them. The schema specification language is a declarative representation language influenced by ideas from knowledge representation (e.g., semantic nets, frames, predicate logic) as well as database schema specification languages and graph data models. The RDF schema specification language is less expressive and simpler to implement than full predicate calculus languages.

RDF adopts a modular approach to metadata that can be considered an implementation of the Warwick Framework. RDF represents an evolution of the Warwick Framework model in that the Warwick Framework allows each metadata vocabulary to be represented in a different syntax. In RDF, all vocabularies are expressed within a single well-defined model. This allows for a finer grained mixing of machine-processable vocabularies and addresses the need to create metadata in which statements can draw upon multiple vocabularies that are managed in a decentralized fashion by independent communities of expertise.

RDF Schemas may be contrasted with XML DTDs and XML Schemas. Unlike an XML DTD or Schema, which gives specific constraints on the structure of an XML document, an RDF Schema provides information about the interpretation of the statements given in an RDF data model. While an XML Schema can be used to validate the syntax of an RDF/XML expression, a syntactic schema alone is not sufficient for RDF purposes. RDF Schemas may also specify constraints that should be followed by these data models.

The RDF Schema specification was directly influenced by consideration of the following problems:

- *Platform for internet content selection (PICS)*: The RDF Model and Syntax is adequate to represent PICS labels; however, it does not provide a general-purpose mapping from PICS rating systems into an RDF representation.
- *Simple web metadata*: An application for RDF is in the description of Web pages. This is one of the basic goals of the Dublin Core Metadata Initiative. The Dublin Core Element Set is a set of 15 elements believed to be broadly applicable to describing Web resources to enable their discovery. The Dublin Core has been a major influence on the development of RDF. An important consideration in the development of the Dublin Core was to not only allow simple descriptions but also to provide the ability to qualify descriptions in order to provide both domain-specific elaboration and descriptive precision.

  The RDF Schema specification provides a machine-understandable system for defining schemas for descriptive vocabularies like the Dublin Core. It allows designers to specify classes of resource types and properties to convey descriptions of those classes, relationships between those properties and classes, and constraints on the allowed combinations of classes, properties, and values.
- *Sitemaps and concept navigation*: A sitemap is a hierarchical description of a Web site. Subject taxonomy is a classification system that might be used by content creators or trusted third parties to organize or classify Web resources. The RDF Schema specification provides a mechanism for defining the vocabularies needed for such applications.

  Thesauri and library classification schemes are examples of hierarchical systems for representing subject taxonomies in terms of the relationships between named concepts. The RDF Schema specification provides sufficient resources for creating RDF models that represent the logical structure of Thesauri and other library classification systems.
- *P3P*: The World Wide Web Consortium (W3C Platform for Privacy Preferences Project (P3P) has specified a grammar for constructing statements about a site's data collection practices and personal preferences as exercised over those practices, as well as a syntax for exchanging structured data.

Although personal data collection practices have been described in P3P using an application-specific XML tagset, there are benefits of using a general metadata model for this data. The structure of P3P policies can be interpreted as an RDF model. Using a metadata schema to describe the semantics of privacy practice descriptions will permit privacy practice data to be used along with other metadata in a query during resource discovery, and will permit a generic software agent to act on privacy metadata using the same techniques as used for other descriptive metadata. Extensions to P3P that describe the specific data elements collected by a site could use RDF Schema to further specify how those data elements are used.

Resources may be instances of one or more classes. Classes are often organized in a hierarchical fashion; for example, a class Cat might be considered a subclass of Mammal, which is a subclass of Animal, meaning that any resource, which is of type Cat, is also considered to be of type Animal. This specification describes a property of a subclass, to denote such relationships between classes.

The RDF Schema type system is similar to the type systems of object-oriented programming languages such as Java. However, RDF differs from many such systems in that instead of defining a class in terms of the properties its instances may have, an RDF schema defines properties in terms of the classes of resource to which they apply. For example, we could define the author property to have a domain of Book and a range of Literal, whereas a classical object-oriented system may typically define a class Book with an attribute called author of type Literal. One benefit of the RDF property-centric approach is that it is very easy for anyone to say anything they want about existing resources, which is one of the architectural principles of the Web.

The following resources are the core classes that are defined as part of the RDF Schema vocabulary. Every RDF model that draws upon the RDF Schema name space (implicitly) includes these:

- *rdfs:Resource*: All things being described by RDF expressions are called resources and are considered to be instances of the class rdfs:Resource. The RDF class rdfs:Resource represents the set called 'Resources' in the formal model for RDF.
- *rdf:Property*: The rdf:Property represents the subset of RDF resources that are properties, that is, all the elements of the set introduced as 'Properties'.
- *rdfs:Class*: This corresponds to the generic concept of a Type or Category, similar to the notion of a Class in object-oriented programming languages such as Java. When a schema defines a new class, the resource representing that class must have an rdf:type property whose value is the resource rdfs:Class. RDF classes can be defined to represent almost anything, such as Web pages, people, document types, databases, or abstract concepts.

Every RDF model that uses the schema mechanism also (implicitly) includes the following core properties. These are instances of the rdf:Property class and provide a mechanism for expressing relationships between classes and their instances or superclasses.

- *rdf:type*: This indicates that a resource is a member of a class, and thus has all the characteristics that are to be expected of a member of that class. When a resource has an rdf:type property whose value is some specific class, we say that the resource is an instance of the specified class. The value of an rdf:type property for some resource is another resource that must be an instance of rdfs:Class. The resource known as rdfs:Class is itself a resource of rdf:type rdfs:Class. Individual classes (e.g., 'Cat') will always have an rdf:type property whose value is rdfs:Class (or some subclass of rdfs:Class).
- *rdfs:subClassOf*: This property specifies a subset/superset relation between classes. The rdfs:subClassOf property is transitive. If class A is a subclass of some broader class B, and B is a subclass of C, then A is also implicitly a subclass of C. Consequently,

resources that are instances of class A will also be instances of C, since A is a subset of both B and C. Only instances of rdfs:Class can have the rdfs:subClassOf property and the property value is always of rdf:type rdfs:Class. A class may be a subclass of more than one class. A class can never be declared to be a subclass of itself, nor of any of its own subclasses.

An example class hierarchy is shown in Figure 7.1. In this figure, we define a class Art. Two subclasses of Art are defined as Painting and Sculpture. We define a class Reproduction – Limited Edition, which is a subclass of both Painting and Sculpture. The arrows in Figure 7.1 point to the subclasses and the type.

RDF schemas can express constraints that relate vocabulary items from multiple independently developed schemas. Since URI references are used to identify classes and properties, it is possible to create new properties whose domain or range constraints reference classes defined in another name space. These constraints include the following:

- The value of a property should be a resource of a designated class. This is a range constraint. For example, a range constraint applying to the author property might express that the value of an author property must be a resource of class Person.
- A property may be used on resources of a certain class. This is a domain constraint. For example, that the author property could only originate from a resource that was an instance of class Book.

RDF uses the XML Name space facility to identify the schema in which the properties and classes are defined. Since changing the logical structure of a schema risks breaking other RDF models that depend on that schema, a new name space URI should be declared whenever an RDF schema is changed.

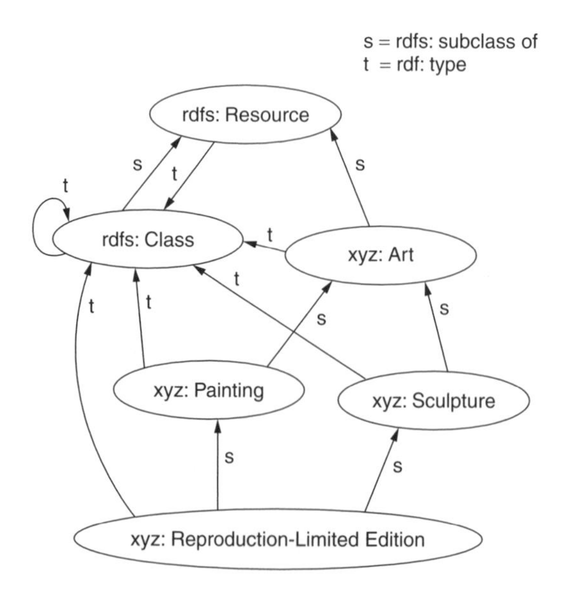

Figure 7.1 Class hierarchy in RDF.

In effect, changing the RDF statements, which constitute a schema, creates a new one; new schema name spaces should have their own URI to avoid ambiguity. Since an RDF Schema URI unambiguously identifies a single version of a schema, software that uses or manages RDF (e.g., caches) should be able to safely store copies of RDF schema models for an indefinite period. The problems of RDF schema evolution share many characteristics with XML DTD version management and the general problem of Web resource versioning.

Since each RDF schema has its own unchanging URI, these can be used to construct unique URI references for the resources defined in a schema. This is achieved by combining the local identifier for a resource with the URI associated with that schema name space. The XML representation of RDF uses the XML name space mechanism for associating elements and attributes with URI references for each vocabulary item used.

The resources defined in RDF schemas are themselves Web resources and can be described in other RDF schemas. This principle provides the basic mechanism for RDF vocabulary evolution. The ability to express specialization relationships between classes (subClassOf) and between properties (subPropertyOf) provides a simple mechanism for making statements about how such resources map to their predecessors. Where the vocabulary defines properties, the same approach can be taken, using rdfs:subPropertyOf to make statements about relationships between properties defined in successive versions of an RDF vocabulary.

7.3 CC/PP – USER SIDE FRAMEWORK FOR CONTENT NEGOTIATION

RDF can be used to create a general, yet extensible framework for describing user preferences and device capabilities. This information can be provided by the user to servers and content providers. The servers can use this information describing the user's preferences to customize the service or content provided. The ability of RDF to reference profile information *via* URLs assists in minimizing the number of network transactions required to adapt content to a device, while the framework fits well into the current and future protocols.

A CC/PP is a collection of the capabilities and preferences associated with user and the agents used by the user to access the World Wide Web. These user agents include the hardware platform, system software, and applications used by the user. User agent capabilities and references can be thought of as metadata or properties and descriptions of the user agent hardware and software.

A description of the user's capabilities and preferences is necessary but insufficient to provide a general content negotiation solution. A general framework for content negotiation requires a means for describing the metadata or attributes and preferences of the user and his/hers/its agents, the attributes of the content and the rules for adapting content to the capabilities and preferences of the user. The mechanisms, such as accept headers and tags, are somewhat limited. For example, the content might be authored in multiple languages with different levels of confidence in the translation and the user might be able to understand multiple languages with different levels of proficiency. To complete the negotiation, some rule is needed for selecting a version of the document on the basis of weighing the user's proficiency in different languages against the quality of the documents various translations.

The CC/PP proposal describes an interoperable encoding for capabilities and preferences of user agents, specifically Web browsers. The proposal is also intended to support applications other than browsers, including e-mail, calendars, and so on. Support for peripherals such as printers and fax machines requires other types of attributes such as type of printer, location, Postscript support, color, and so on. We believe an XML/RDF-based approach would be suitable. However, metadata descriptions of devices such as printers or fax machines may use a different scheme.

The basic data model for a CC/PP is a collection of tables. Though RDF makes modeling a wide range of data structures possible, it is unlikely that this flexibility will be used in the creation of complex data models for profiles. In the simplest form, each table in the CC/PP is a collection of RDF statements with simple, atomic properties. These tables may be constructed from default settings, persistent local changes, or temporary changes made by a user. One extension to the simple table of properties data model is the notion of a separate, subordinate collection of default properties. Default settings might be properties defined by the vendor. In the case of hardware, the vendor often has a very good idea of the physical properties of any given model of product. However, the current owner of the product may be able to add options, such as memory or persistent store or additional I/O devices that add new properties or change the values of some original properties. These would be persistent local changes. An example of a temporary change would be turning sound on or off.

The profile is associated with the current network session or transaction. Each major component may have a collection of attributes or preferences. Examples of major components are the hardware platform, upon which all the software is executing, the software platform, upon which all the applications are hosted, and each of the applications.

Some collections of properties and property values may be common to a particular component. For example, a specific model of a smart phone may come with a specific CPU, screen size, and amount of memory by default. Gathering these default properties together as a distinct RDF resource makes it possible to independently retrieve and cache those properties. A collection of default properties is not mandatory, but it may improve network performance, especially the performance of relatively slow wireless networks.

From the point of view of any particular network transaction, the only property or capability information that is important is whatever is current. The network transaction does not care about the differences between defaults or persistent local changes; it only cares about the capabilities and preferences that apply to the current network transaction. Because this information may originate from multiple sources and because different parts of the capability profile may be differentially cached, the various components must be explicitly described in the network transaction.

The CC/PP is the encoding of profile information that needs to be shared between a client and a server, gateway, or proxy. The persistent encoding of profile information and the encoding for the purposes of interoperability (communication) need not be the same.

Instead of enumerating each set of attributes, a remote reference can be used to name a collection of attributes such as the hardware platform defaults. This has the advantage of enabling the separate fetching and caching of functional subsets. This might be very good if the link between the gateway or the proxy and the client agent was slow and the link between the gateway or proxy and the site named by the remote reference was fast – a typical case when the user agent is a smart phone. Another advantage is the simplification of the development of different vocabularies for hardware vendors and software vendors.

It is important to be able to add to and modify attributes associated with the current CC/PP. We need to be able to modify the value of certain attributes, such as turning sound on and off and we need to make persistent changes to reflect things like a memory upgrade. We need to be able to override the default profile provided by the vendor.

When used in the context of a Web-browsing application, a CC/PP should be associated with a notion of a current session rather than a user or a node. HTTP and WSP (the WAP session protocol) both define different session semantics. The client, server, gateways, and proxies may already have their own, well-defined notions of what constitutes a connection or a session. The protocol strategy is to send as little information as possible and if anyone is missing something, they have to ask for it. If there is good reason to believe that someone is going to ask for a profile, the client can elect to send the most efficient form of the profile that makes sense.

We consider the following possible interaction between a server and a client. When the client begins a session, it sends a minimal profile using as much indirection as possible. If the server/gateway/proxy does not have a CC/PP for this session, then it asks for one. When a profile is sent, the client tries a minimal form, that is, it uses as much indirection as possible and only names the nondefault attributes of the profile. The server/gateway/proxy can try to fill in the profile using the indirect HTTP references (which may be independently cached). If any of these fail, a request for additional data can be sent to the user, which can reply with a fully enumerated profile. If the client changes the value of an attribute, such as turning sound off, only that change needs to be sent.

It is likely that servers and gateways/proxies are concerned with different preferences. For example, the server may need to know which language the user prefers and the gateway may have responsibility to trim images to eight bits of color (to save bandwidth). However, the exact use of profile information by each server/gateway/proxy is hard to predict. Therefore, gateways/proxies should forward all profile information to the server. Any requests for profile information that the gateway/proxy cannot satisfy should be forwarded to the client.

The ability to compose a profile from sources provided by third parties at run-time exposes the system to a new type of attack. For example, if the URL that named the hardware default platform defaults were to be compromised *via* an attack on domain name system (DNS), it would be possible to load incorrect profile information. If cached within a server/gateway/proxy, this could be a serious denial of service attack. If this is a serious enough problem, it may be worth adding digital signatures to the URLs used to refer to profile components.

The CC/PP framework is a mechanism for describing the capabilities and preferences associated with users and user agents accessing the World Wide Web. Information about user agents includes the hardware platform, system software, applications, and user preferences. The user agent capabilities and preferences can be thought of as metadata, or properties and descriptions of the user agent's hardware and software. The CC/PP descriptions are intended to provide information necessary to adapt the content and the content delivery mechanisms to best fit the capabilities and preferences of the user and its agents.

The major disadvantage of this format is that it is verbose. Some networks are very slow and this would be a moderately expensive way to handle metadata. There are several optimizations possible to help deal with network performance issues. One strategy is to use a compressed form of XML, and a complementary strategy is to use references (URIs). Instead of enumerating each set of attributes, a reference can be used to name a collection of attributes such as the hardware platform defaults. This has the advantage of enabling the separate fetching and caching of functional subsets.

Another problem is to propagate changes to the current CC/PP descriptions to an origin server, a gateway, or a proxy. One solution is to transmit the entire CC/PP descriptions with each change. This is not ideal for slow networks. An alternative is to send only the changes.

The CC/PP exchange protocol does not depend on the profile format that it conveys. Therefore, another profile format besides the CC/PP description format can be applied to the CC/PP exchange protocol.

The basic requirements for the CC/PP exchange protocol are as follows:

- The transmissions of the CC/PP descriptions should be HTTP/1.1-compatible.
- The CC/PP exchange protocol should support an indirect addressing scheme based on Request For Comment RFC2396 (Generic Syntax for URIs) for referencing profile information.
- Components used to construct CC/PP descriptions, such as vendor default descriptions, should be independently cacheable.
- The CC/PP exchange protocol should provide a lightweight exchange mechanism that permits the client to avoid resending the elements of the CC/PP descriptions that have not changed since the last time the information was transmitted.

CC/PP repository is an application program that maintains CC/PP descriptions. The CC/PP repository should be HTTP/1.0 or HTTP/1.1-compliant. The CC/PP repository is not required to comply with the CC/PP exchange protocol.

The protocol strategy is to send a request with profile information, which is as limited as possible, by using references (URIs). For example, a user agent issues a request with URIs that address the profile information, and if the user agent changes the value of an attribute, such as turning sound off, only that change is sent together with the URIs. When an origin server receives the request, the origin server inquires of CC/PP repositories the CC/PP descriptions using the list of URIs. Then the origin server creates a tailored content using the fully enumerated CC/PP descriptions.

The origin server might not obtain the fully enumerated CC/PP descriptions when any one of the CC/PP repositories is not available. In this case, it depends on the implementation whether the origin server should respond to the request with a tailored content, a nontailored content, or an error. In any case, the origin server should inform the user agent of the fact. A warning mechanism is introduced for this purpose.

It is likely that an origin server, a gateway, or a proxy will be concerned with different device capabilities or user preferences. For example, the origin server may have responsibility to select content according to the user-preferred language, while the proxy may have responsibility to transform the encoding format of the content. Therefore, gateways or proxies might not forward all profile information to an origin server.

The CC/PP exchange protocol might convey natural language codes within header field values. Therefore, internationalization issues must be considered. The internationalization policy of the CC/PP exchange protocol is based on RFC2277 (IETF Policy on Character Sets and Language).

Considering how to maintain a session like real-time streaming protocol (RTSP) is worthwhile from the point of view of minimizing transactions (i.e., the session mechanism could permit the client to avoid resending the elements of the CC/PP descriptions that have not changed since the last time the information was transmitted). However, a session mechanism would reduce cache efficiency and requires maintaining states between a user agent and an origin server. The CC/PP exchange protocol is designed as a session-less (stateless) protocol.

The CC/PP exchange protocol is based on the HTTP Extension Framework. The HTTP Extension Framework is a generic extension mechanism for HTTP/1.1, which is designed to interoperate with existing HTTP applications.

An extension declaration is used to indicate that an extension has been applied to a message and possibly to reserve a part of the header name space identified by a header field prefix. The HTTP Extension Framework introduces two types of extension declaration strength: mandatory and optional, and two types of extension declaration scope: hop-by-hop and end-to-end. Which type of the extension declaration strengths and/or which type of the extension declaration scopes should be used depends on what the user agent needs to do.

The strength of the extension declaration should be mandatory if the user agent needs to obtain an error response when a server (an origin server, a gateway, or a proxy) does not comply with the CC/PP exchange protocol. The strength of the extension declaration should be optional if the user agent needs to obtain the nontailored content when a server does not comply with the CC/PP exchange protocol.

The scope of the extension declaration should be hop-by-hop if the user agent has an *a priori* knowledge that the first-hop proxy complies with the CC/PP exchange protocol. The scope of the extension declaration should be end-to-end if the user agent has an *a priori* knowledge that the first-hop proxy does not comply with the CC/PP exchange protocol, or the user agent does not use a proxy. The integrity and persistence of the extension should be maintained and kept unquestioned throughout the lifetime of the extension. The name space prefix is generated dynamically.

The profile header field is a request-header field, which conveys a list of references that address CC/PP descriptions. The goal of the CC/PP framework is to specify how client devices express their capabilities and preferences (the user agent profile) to the server that originates content (the origin server). The origin server uses the user agent profile to produce and deliver content appropriate to the client device. In addition to computer-based client devices, particular attention is paid to other kinds of devices such as mobile phones.

The requirements on the framework emphasize three aspects: flexibility, extensibility, and distribution. The framework must be flexible, since we cannot today predict all the different types of devices that will be used in the future, or the ways those devices will be used. It must be extensible for the same reasons: it should not be hard to add and test new descriptions. And it must be distributed, since relying on a central registry might make it inflexible.

The basic problem that the CC/PP framework addresses is to create a structured and universal format for how a client device tells an origin server about its user agent profile. A design used to convey the profile is independent on the protocols used to transport it. It does not present mechanisms or protocols to facilitate the transmission of the profile.

The framework describes a standardized set of CC/PP attributes – a vocabulary – that can be used to express a user agent profile in terms of capabilities and the users preferences for the use of these capabilities. This is implemented using the XML application RDF. This enables the framework to be flexible, extensible, and decentralized, thus fulfilling the requirements.

RDF is used to express the client device's user agent profile. The client device may be a workstation, personal computer, mobile terminal, or set-top box. When used in a request-response protocol like HTTP, the user agent profile is sent to the origin server that, subsequently, produces content that satisfies the constraints and preferences expressed in the user agent profile. The CC/PP framework may be used to convey to the client device what variations in the requested content are available from the origin server.

Fundamentally, the CC/PP framework starts with RDF and then overlays a CC/PP-defined set of semantics that describe profiles. The CC/PP framework does not specify whether the client device or the origin server initiates this exchange of profiles. The CC/PP framework specifies the RDF usage and associated semantics that should be applied to all profiles that are being exchanged.

The HTTP use case with repository for the profile information is as follows:

1. Request from client with profile information
2. Server resolves and retrieves profile (from CC/PP repository in the network), and uses it to adapt the content
3. Server returns adapted content
4. Proxy forwards response to the client.

The notion of a proxy resolving the information and retrieving it from a repository might assume the use of an XML processor and encoding of the profile in XML.

In case the document contains a profile, the above could still apply. However, there will be some interactions inside the server, as the client profile information needs to be matched with the document profile. The interactions in the server are not defined.

The document profile use case is as follows:

1. Request (extended method) with profile information
2. Document profile is matched against device profile to derive optimum representation

3. Document is adapted
4. Response to the client with adapted content.

The mobile environment requires small messages and has a much narrower bandwidth than fixed environments.

When a user agent profile is used with a WAP device, the scenario is as follows:

1. WSP request with profile information or difference relative to a specified default.
2. Gateway caches WSP header, composes the current profile (using the cached header as defaults and diffs from the client). The user agent profile values can change at setup or resume of session.
3. Gateway passes request to server using extended HTTP method.
4. Server returns adapted information.
5. Response in WSP with adapted content.

The user agent profile is transmitted as a parameter of the WSP session to the WAP gateway and cached; it is then transferred over HTTP using the CC/PP Exchange Protocol, which is an application of the HTTP Extension Framework.

The WAP system uses wireless markup language (WML) as its content format, not HTML. This is an XML application, and the adaptation could, for instance, be transformation from another XML format into WML.

The Conneg (Content Negotiation) working group in the IETF has developed a form of media feature descriptors, which are registered with Internet Assigned Numbers Authority (IANA). Like the CC/PP format and vocabulary, this is intended to be independent of protocol. The Conneg working group also defined a matching semantics based on constraints.

The Conneg framework defines an IANA registry for feature tags, which are used to label media feature values associated with the presentation of data (e.g., display resolution, color capabilities, audio capabilities, etc.). To describe a profile, Conneg uses predicate expressions (feature predicates) on collections of media feature values (feature collection) as an acceptable set of media feature combinations (feature set). The same basic framework is applied to describe receiver and sender capabilities and preferences, and also document characteristics. Profile matching is performed by finding the feature set that matches two (or more) profiles. This involves finding the feature predicate that is equivalent to the logical-AND of the predicates being matched.

Conneg is protocol independent, but can be used for server-initiated transactions, for example:

1. Server sends to proxy
2. Proxy retrieves profile from client (or checks against a cache)
3. Client returns profile
4. Proxy formats information and forwards it.

The TV/broadcast use case describes a push situation, in which a broadcaster sends out an information set to a device without a back channel. The server cannot get capabilities for all devices, so it broadcasts a minimum set of elements or a multipart document, which is then adapted to the optimal presentation for the device. Television manufacturers desire to turn their appliances into interactive devices. This effort is based on the use of extensible HTML (XHTML) as language for the content representation, which, for instance, enables the use of content profiles as seen. A television set does not have a local intelligence of its own and does not allow for bidirectional communication with the origin server. This architecture also applies to several different device classes, such as pagers, e-mail clients, and other similar devices. It is not the case that they are entirely without interaction, however. In reality, these devices follow a split-client model, in which the broadcaster, cable head-end, or similar entity interacts with the origin server and sends a renderable version of the content to the part of the client, which resides at the user site.

There are also use cases in which the entire data set is downloaded into the client, and the optimal rendering is constructed there, for instance, in a set-top box. In these cases, the CC/PP client profile will need to be matched against a document profile representing the author's preferences for the rendering of the document.

The protocol interactions are as follows:

1. Document is pushed to the client including alternate information and document profile.
2. Client matches the rules in the document profile and its own profile.
3. The client adapts content to its optimal presentation using the derived intersection of the two sets.

When a request for content is made by a user agent to an origin server, a CC/PP profile describing the capabilities and preferences is transmitted along with the request. It is possible that intermediate network elements such as gateways and transcoding proxies that have additional capabilities might be able to translate or adapt the content before rendering it to the device. Such capabilities are not known to the user agents and therefore cannot be included in the original profile. However, these capabilities would need to be conveyed to the origin server or proxy serving/generating the content. In some instances, the profile information provided by the requesting client device may need to be overridden or augmented.

CC/PP framework must therefore support the ability for such proxies and gateways to assert their capabilities using the existing vocabulary or extensions thereof. This can be done as amendments or overrides to the profile included in the request. Given the use of XML as the base format, these can be in-line references to be downloaded from a repository as the profile is resolved.

The protocol interactions are as follows:

1. The CC/PP-compliant user agent requests content with the profile.
2. The transcoding proxy appends additional capabilities (profile segment), or overrides the default values, and forwards the profile to the network.
3. The origin server constructs the profile and generates adapted content.
4. The transcoding proxy transcodes the content received on the basis of its abilities, and forwards the resulting customized content to the device for rendering.

The foundation of RDF is a model for representing named properties and property values. The RDF model draws on principles from various data representation communities.

RDF properties may be thought of as attributes of resources and in this sense correspond to traditional attribute-value pairs. RDF properties also represent relationships between resources and an RDF model can therefore resemble an entity-relationship diagram. In object-oriented design terminology, resources correspond to objects and properties correspond to instance variables.

The RDF data model is a syntax-neutral way of representing RDF expressions. The data model representation is used to evaluate equivalence in meaning. Two RDF expressions are equivalent if and only if their data model representations are the same. This definition of equivalence permits some syntactic variation in expression without altering the meaning.

The basic data model consists of three object types:

- *Resources*: Resources are described by RDF expressions. A resource may be an entire Web page, a part of a Web page, for example, a specific HTML or XML element within the document source. A resource may also be a whole collection of pages, for example, an entire Web site. A resource may also be an object that is not directly accessible *via* the Web, for example, a printed book. Anything can have a URI; the extensibility of URIs allows the introduction of identifiers for any entity.
- *Properties*: A property is a specific aspect, characteristic, attribute, or relation used to describe a resource. Each property has a specific meaning, defines its permitted values, the types of resources it can describe, and its relationship with other properties.
- *Statements*: A specific resource together with a named property plus the value of that property for that resource is an RDF statement. These three individual parts of a statement are called the subject, the predicate, and the object, respectively. The object of a statement (i.e., the property value) can be another resource or it can be a literal, that is, a resource (specified by a URI) or a simple string or other primitive datatype defined by XML. In RDF terms, a literal may have content that is XML markup but is not further evaluated by the RDF processor. There are some syntactic restrictions on how markup in literals may be expressed.

RDF properties may be thought of as attributes of resources and in this sense correspond to traditional attribute-value pairs. RDF properties also represent relationships between resources. As such, the RDF data model can therefore resemble an entity-relationship diagram. The RDF data model, however, provides no mechanisms for declaring these properties, nor does it provide any mechanisms for defining the relationships between these properties and other resources. That is the role of RDF Schema.

Each RDF schema is identified by its own static URI. The schema's URI can be used to construct unique URI references for the resources defined in a schema. This is achieved by combining the local identifier for a resource with the URI associated with that schema name space. The XML representation of RDF uses the XML name space mechanism for associating elements and attributes with URI references for each vocabulary item used.

A CC/PP profile describes client capabilities in terms of a number of CC/PP attributes or features. Each of these features is identified by a name in the form of a URI. A collection of such names used to describe a client is called a vocabulary.

CC/PP defines a small, core set of features that are applicable to a wide range of user agents and that provide a broad indication of a clients capabilities. This is called the core vocabulary. It is expected that any CC/PP processor will recognize all the names in the core vocabulary, together with an arbitrary number of additional names drawn from one or more extension vocabularies.

When using names from the core vocabulary or an extension vocabulary, it is important that all system components (clients, servers, proxies, etc.), which generate or interpret the names, apply a common meaning to the same name. It is preferable that different components use the same name to refer to the same feature, even when they are a part of different applications, as this improves the chances of effective interworking across applications that use capability information.

Within an RDF expression describing a device, a vocabulary name appears as the label on a graph edge linking a resource to a value for the named attribute. The attribute value may be a simple string value, or another resource, with its own attributes representing the component parts of a composite value.

Vocabulary extensions are used to identify more detailed information than can be described using the core vocabulary. Any application or operational environment that uses CC/PP may define its own vocabulary extensions, but wider interoperability is enhanced if vocabulary extensions are defined, which can be used more generally, for example, a standard extension vocabulary for imaging devices, or voice messaging devices, or wireless access devices, and so on.

Any CC/PP expression can use terms drawn from an arbitrary number of different vocabularies, so there is no restriction caused by reusing terms from an existing vocabulary rather then defining new names to identify the same information.

CC/PP attribute names are in the form of a URI. Any CC/PP vocabulary is associated with an XML name space, which combines a base URI with a local XML element name (or XML attribute name) to yield a URI corresponding to an element name. Thus, CC/PP vocabulary terms are constructed from an XML name space base URI and a local attribute name.

Anyone can define and publish a CC/PP vocabulary extension (assuming administrative control or allocation of a URI for an XML name space). For such a vocabulary to be useful, it must be interpreted in the same way by communicating entities. Thus, use of an existing extension vocabulary or publication of a new vocabulary definition containing detailed descriptions of the various CC/PP attribute names is encouraged wherever possible. Many extension vocabularies will be drawn from existing applications and protocols.

CC/PP expresses the user agent capabilities and how the user wants to use them. XHTML document profiles express the required functionalities for what the author perceives as optimal rendering and how the author wants them to be used. We regard the CC/PP format as the common format, to which other profile formats have been mapped. The interactions are as follows:

1. Request (extended method) with profile information.
2. Profile translation (this refers to functional elements. The entire process can also take place in the origin server).
3. Schema for document profile is retrieved (from a repository or other entity).
4. Server resolves mappings and creates an intermediary CC/PP schema for the matching.
5. Document profile is matched against device profile to derive optimum representation.

6. Document is adapted.
7. Response to client with adapted content. Depending on the format of the document profile, the translation can be done in different ways.
8. In the case of a dedicated XML-based format, mapping the XML Schema for the dedicated format to the schema for RDF will allow the profile to be expressed as RDF by the translating entity. In the case of a non-XML-based format, a one-to-one mapping will have to be provided for the translation.

7.4 CC/PP EXCHANGE PROTOCOL BASED ON THE HTTP EXTENSION FRAMEWORK

The CC/PP framework is a mechanism for describing the capabilities and preferences associated with users and user agents accessing the World Wide Web. Information about user agents includes the hardware platform, system software, applications, and user preferences (P3P). The user agent capabilities and preferences can be thought of as metadata, or properties and descriptions of the user agent's hardware and software. The CC/PP descriptions are intended to provide information necessary to adapt the content and the content delivery mechanisms to best fit the capabilities and preferences of the user and its agents.

Instead of enumerating each set of attributes, a reference can be used to name a collection of attributes such as the hardware platform defaults. This has the advantage of enabling the separate fetching and caching of functional subsets.

Another problem is to propagate changes to the current CC/PP descriptions to an origin server, a gateway, or a proxy. One solution is to transmit the entire CC/PP descriptions with each change. This is not ideal for slow networks. An alternative is to send only the changes.

The CC/PP exchange protocol does not depend on the profile format that it conveys. Therefore, another profile format besides the CC/PP description format can be applied to the CC/PP exchange protocol.

The basic requirements for the CC/PP exchange protocol are as follows:

1. The transmissions of the CC/PP descriptions should be HTTP/1.1-compatible.
2. The CC/PP exchange protocol should support an indirect addressing scheme based on RFC2396 for referencing profile information.
3. Components used to construct CC/PP descriptions, such as vendor default descriptions, should be independently cacheable.
4. The CC/PP exchange protocol should provide a lightweight exchange mechanism that permits the client to avoid resending the elements of the CC/PP descriptions that have not changed since the last time the information was transmitted.

For example, a user agent issues a request with URIs that address the profile information, and if the user agent changes the value of an attribute, such as turning sound off, only that change is sent together with the URIs. When an origin server receives the request, the origin server inquires of CC/PP repositories the CC/PP descriptions using the list of URIs. Then the origin server creates a tailored content using the fully enumerated CC/PP descriptions.

The origin server might not obtain the fully enumerated CC/PP descriptions when any one of the CC/PP repositories is not available. In this case, it depends on the implementation whether the origin server should respond to the request with a tailored content, a nontailored content, or an error. In any case, the origin server should inform the user agent of the fact. A warning mechanism is introduced for this purpose.

It is likely that an origin server, a gateway, or a proxy will be concerned with different device capabilities or user preferences. For example, the origin server may have responsibility to select content according to the user-preferred language, while the proxy may have responsibility to transform the encoding format of the content. Therefore, gateways or proxies might not forward all profile information to an origin server.

The CC/PP exchange protocol is based on the HTTP Extension Framework. The HTTP Extension Framework is a generic extension mechanism for HTTP/1.1, which is designed to interoperate with existing HTTP applications.

An extension declaration is used to indicate that an extension has been applied to a message and possibly to reserve a part of the header name space identified by a header field prefix. The HTTP Extension Framework introduces two types of extension declaration strength: mandatory and optional, and two types of extension declaration scope: hop-by-hop and end-to-end.

Which type of the extension declaration strengths and/or which type of the extension declaration scopes should be used depends on what the user agent needs to do.

The strength of the extension declaration should be mandatory if the user agent needs to obtain an error response when a server (an origin server, a gateway, or a proxy) does not comply with the CC/PP exchange protocol. The strength of the extension declaration should be optional if the user agent needs to obtain the nontailored content when a server does not comply with the CC/PP exchange protocol.

The scope of the extension declaration should be hop-by-hop if the user agent has an *a priori* knowledge that the first-hop proxy complies with the CC/PP exchange protocol. The scope of the extension declaration should be end-to-end if the user agent has an *a priori* knowledge that the first-hop proxy does not comply with the CC/PP exchange protocol or the user agent does not use a proxy.

The absoluteURI in the Profile header field addresses an entity of a CC/PP description, which exists in the World Wide Web. CC/PP descriptions may originate from multiple sources (e.g., hardware vendors, software vendors, etc). A CC/PP description that is provided by a hardware vendor or a software vendor should be addressed by an absoluteURI. A user agent issues a request with these absoluteURIs in the Profile header instead of sending whole CC/PP descriptions, which contributes to reducing the amount of transaction. The syntax of the absoluteURI must conform to RFC2396.

The scenario of mandatory and end-to-end using the CC/PP exchange protocol is as follows:

1. The user agent issues a mandatory extension request.
2. The origin server examines the extension declaration header and determines if it is supported for this message, if not, it responds with not extended status code.

3. Otherwise, the origin server gets the list of the references in the Profile header field.
4. The origin server generates the tailored content according to the enumerated CC/PP descriptions and sends back the tailored content with the mandatory extension response header.

In this example, the content is not cacheable so that the origin server indicates no-cache directives in the Cache-control header field.

The scenario of the optional and end-to-end using the CC/PP exchange protocol is as follows:

1. The user agent issues an optional extension request.
2. The origin server examines the extension declaration header and determines if it is supported for this message. If not, the origin server ignores the extension headers and sends back the nontailored content.
3. Otherwise, the origin server gets the list of the absoluteURIs in the Profile header field. After that, the origin server issues requests to the CC/PP repositories to get the CC/PP descriptions using these absoluteURIs.
4. The origin server generates the tailored content according to the enumerated CC/PP descriptions and sends back the tailored content.

The scenario of the mandatory and hop-by-hop using CC/PP exchange protocol is as follows:

1. The user agent issues a mandatory extension request.
2. The first-hop proxy examines the extension declaration header and determines if it is supported for this message. If not, it responds with a not extended status code.
3. Otherwise, the first-hop proxy issues requests to the CC/PP repositories to get the CC/PP descriptions using the absoluteURIs.
4. The first-hop proxy generates the request with the Accept, Accept-Charset, Accept-Encoding, and Accept-Language, using the enumerated CC/PP descriptions, and issues the request to the origin server.
5. The origin server responds to the first-hop proxy with the content.
6. The first-hop proxy transforms the content into the tailored content using the enumerated CC/PP descriptions. After that, the first-hop proxy sends back the tailored content with the mandatory hop-by-hop extension response header.

The scenario of the optional and hop-by-hop by using CC/PP exchange protocol is as follows:

1. The user agent issues an optional extension request.
2. The first-hop proxy examines the extension declaration header and determines if it is supported for this message. If not, the first-hop proxy forwards requests to the origin server after the first-hop proxy removes the headers that are listed in the Connection header.
3. Otherwise, the first-hop proxy issues requests to the CC/PP repositories to get the CC/PP descriptions using the absoluteURIs.
4. The first-hop proxy generates the request and issues the request to the origin server.

5. The origin server responds to the first-hop proxy with the content.
6. The first-hop proxy transforms the content into the tailored content using the enumerated CC/PP descriptions. After that, the first-hop proxy sends back the tailored content to the user agent.

The scenario of the response with warning using the CC/PP exchange protocol is as follows:

1. The user agent issues a request.
2. The origin server issues requests to the CC/PP repositories to get the CC/PP descriptions.
3. The CC/PP description is obtained successfully from or the CC/PP description could not be obtained.
4. The origin server generates the tailored content using only the CC/PP description obtained successfully and sends back the tailored content with the Profile-Warning response header. (When the origin server did not obtain the fully enumerated CC/PP descriptions, it depends on the implementation whether the origin server should respond to the request with a tailored content, a nontailored content, or an error.)

The scenario how to enable the HTTP cache expiration model (end-to-end) using CC/PP exchange protocol is as follows:

1. The user agent issues a request.
2. The origin server issues requests to the CC/PP repositories to get the CC/PP descriptions.
3. The origin server generates and sends back the tailored content.

The scenario how to enable the HTTP cache expiration model (hop-by-hop) using the CC/PP exchange protocol is as follows:

1. The user agent issues a request.
2. The first-hop proxy issues requests to the CC/PP repositories to get the CC/PP descriptions.
3. The first-hop proxy generates and issues a request to the origin server.
4. The origin server responds to the first-hop proxy with the content.
5. The first-hop proxy transforms and sends back a tailored content with the Cache-control header, the Vary header, and the Expires header. Therefore the response might be used by the user agent without revalidation.

7.5 REQUIREMENTS FOR A CC/PP FRAMEWORK, AND THE ARCHITECTURE

The goal of the CC/PP framework is to specify how client devices express their capabilities and preferences (the user agent profile) to the server that originates content (the origin server). The origin server uses the user agent profile to produce and deliver content appropriate to the client device. In addition to computer-based client devices, particular attention is paid to other kinds of devices such as mobile phones.

The requirements on the framework emphasize three aspects: flexibility, extensibility, and distribution. The framework must be flexible, since we cannot today predict all the different types of devices that will be used in the future, or the ways in which those devices will be used. It must be extensible for the same reasons: it should not be hard to add and test new descriptions; and it must be distributed, since relying on a central registry might make it inflexible.

The basic problem, which the CC/PP framework addresses, is to create a structured and universal format for how a client device tells an origin server about its user agent profile. We present a design that can be used to convey the profile and is independent on the protocols used to transport it. It does not present mechanisms or protocols to facilitate the transmission of the profile.

The framework describes a standardized set of CC/PP attributes, a vocabulary that can be used to express a user agent profile in terms of capabilities and the users preferences for the use of these capabilities. This is implemented using the XML application RDF. This enables the framework to be flexible, extensible, and decentralized, thus fulfilling the requirements.

RDF is used to express the client device's user agent profile. The client device may be a workstation, personal computer, mobile terminal, or set-top box.

When used in a request-response protocol like HTTP, the user agent profile is sent to the origin server, which, subsequently, produces content that satisfies the constraints and preferences expressed in the user agent profile. The CC/PP framework may be used to convey to the client device what variations in the requested content are available from the origin server.

Fundamentally, the CC/PP framework starts with RDF and then overlays a CC/PP-defined set of semantics that describe profiles. The CC/PP framework does not specify whether the client device or the origin server initiates this exchange of profiles. The CC/PP framework specifies the RDF usage and associated semantics that should be applied to all profiles that are being exchanged.

Using the World Wide Web with content negotiation as it is designed today enables the selection of a variant of a document. Using an extended capabilities description, an optimized presentation can be produced. This can take place by selecting a style sheet that is transmitted to the client or by selecting a style sheet that is used for transformations. It can also take place through the generation of content or transformation.

The CC/PP Exchange Protocol extends this model by allowing for the transmission and caching of profiles and the handling of profile differences. This use case in itself consists of two different use cases: the origin server receives the CC/PP profile directly from the client; and the origin server retrieves the CC/PP profile from an intermediate repository.

In this case, the profile is used by an origin server on the Web to adapt the information returned in the request. In the HTTP use case, when the interaction passes directly between a client and a server, the user agent sends the profile information with the request and the server returns adapted information. The interaction takes place over an extended HTTP method.

When the profile is composed by resolving in-line references from a repository for the profile information, the process is as follows:

1. Request from client with profile information
2. Server resolves and retrieves profile (from CC/PP repository on the network), and uses it to adapt the content
3. Server returns adapted content
4. Proxy forwards response to client.

The notion of a proxy resolving the information and retrieving it from a repository may assume the use of an XML processor and encoding of the profile in XML.

In case the document contains a profile, there will be some interactions inside the server, as the client profile information needs to be matched with the document profile. The interactions in the server are not defined.

The document profile use case is as follows:

1. Request (extended method) with profile information.
2. Document profile is matched against device profile to derive optimum representation.
3. Document is adapted.
4. Response to the client with adapted content.

The requirement is that the integrity of the information is guaranteed during transit. In the proxy use case, a requirement is the existence of a method to resolve references in the proxy. This might presume the use of an XML processor and XML encoding. The privacy of the user needs to be safeguarded. The document profile and the device profile can use a common vocabulary for common features. They can also use compatible feature constraining forms so that it is possible to match a document profile against a receiver profile and determine compatibility. If not, a mapping needs to be provided for the matching to take place.

The WAP Forum architecture is based on a proxy server, which acts as a gateway to the optimized protocol stack for the mobile environment. It is to this proxy that the mobile device connects. On the wireless side of the communication, it uses an optimized, stateful protocol (Wireless Session Protocol, WSP; and an optimized transmission protocol, Wireless Transaction Protocol, WTP); on the fixed side of the connection, it uses HTTP. The content is marked up in WML, the WML of the WAP Forum. The mobile environment requires small messages and has a much narrower bandwidth than fixed environments.

When a user agent profile is used with a WAP device, it performs as follows:

1. WSP request with profile information or difference relative to a specified default
2. Gateway caches WSP header and composes the current profile
3. Gateway passes request to server using extended HTTP method
4. Server returns adapted information
5. Response in WSP with adapted content.

The user agent profile is transmitted as a parameter of the WSP session to the WAP gateway and cached; it is then transferred over HTTP using the CC/PP Exchange Protocol, which is an application of the HTTP Extension Framework. The WAP system uses WML as its content format, not HTML. This is an XML application, and the adaptation could, for instance, be transformation from another XML format into WML.

7.6 SUMMARY

XML documents are made up of storage units called entities, which contain either parsed or unparsed data. Parsed data is made up of characters, some of which form character data and some of which form markup. Markup encodes a description of the document's storage layout and logical structure. XML provides a mechanism to impose constraints on the storage layout and logical structure.

The RDF is a foundation for processing metadata; it provides interoperability between applications that exchange machine-understandable information on the Web. RDF uses XML to exchange descriptions of Web resources but the resources being described can be of any type, including XML and non-XML resources. RDF emphasizes facilities to enable automated processing of Web resources. RDF can be used in a variety of application areas, for example, in resource discovery to provide better search engine capabilities; in cataloging for describing the content and content relationships available at a particular Web site, page, or digital library, by intelligent software agents to facilitate knowledge sharing and exchange; in content rating; in describing collections of pages that represent a single logical document; in describing intellectual property rights of Web pages; and in expressing the privacy preferences of a user as well as the privacy policies of a Web site. RDF with digital signatures is the key to building the Web of Trust for electronic commerce, collaboration, and other applications.

The goal of the CC/PP framework is to specify how client devices express their capabilities and preferences (the user agent profile) to the server that originates content (the origin server). The origin server uses the user agent profile to produce and deliver content appropriate to the client device. In addition to computer-based client devices, particular attention is paid to other kinds of devices such as mobile phones.

The requirements on the framework emphasize three aspects: flexibility, extensibility, and distribution. The framework must be flexible since we cannot today predict all the different types of devices that will be used in the future or the ways those devices will be used. It must be extensible for the same reasons: it should not be hard to add and test new descriptions, and it must be distributed, since relying on a central registry might make it inflexible.

PROBLEMS TO CHAPTER 7

XML, RDF, and CC/PP

Learning objectives

After completing this chapter, you are able to

- demonstrate an understanding of XML,
- explain the role of RDF,
- explain the role of CC/PP.

Practice problems

7.1: What is XML?
7.2: What is the role of RDF?
7.3: What is the role of CC/PP?

Practice problem solutions

7.1: XML describes a class of data objects called XML documents and partially describes the behavior of the computer programs that process them. XML is an application profile or restricted form of the SGML.

XML documents are made up of storage units called entities, which contain either parsed or unparsed data. Parsed data is made up of characters, some of which form character data and some of which form markup. Markup encodes a description of the document's storage layout and logical structure. XML provides a mechanism to impose constraints on the storage layout and logical structure.

A software module called an XML processor is used to read XML documents and provide access to their content and structure. It is assumed that an XML processor is doing its work on behalf of another module called the application. XML processor reads XML data and provides the information to the application.

7.2: The RDF is a foundation for processing metadata; it provides interoperability between applications that exchange machine-understandable information on the Web. RDF uses XML to exchange descriptions of Web resources but the resources being described can be of any type, including XML and non-XML resources. RDF emphasizes facilities to enable automated processing of Web resources. RDF can be used in a variety of application areas, for example, in resource discovery to provide better search engine capabilities; in cataloging for describing the content and content relationships available at a particular Web site, page, or digital library, by intelligent software agents to facilitate knowledge sharing and exchange; in content rating; in describing collections of pages that represent a single logical document; in describing intellectual property rights of Web pages; and in expressing the privacy preferences of a user as well as the privacy policies of a Web site. RDF with digital signatures is the key to building the Web of Trust for electronic commerce, collaboration, and other applications.

RDF can be used to create a general, yet extensible framework for describing user preferences and device capabilities. This information can be provided by the user to servers and content providers. The servers can use this information describing the user's preferences to customize the service or content provided. The ability of RDF to reference profile information *via* URLs assists in minimizing the number of network transactions required to adapt content to a device, while the framework fits well into the current and future protocols.

7.3: A CC/PP is a collection of the capabilities and preferences associated with user and the agents used by the user to access the World Wide Web. These user agents include the hardware platform, system software, and applications used by the user. User agent capabilities and references can be thought of as metadata or properties and descriptions of the user agent hardware and software.

The basic data model for a CC/PP is a collection of tables. Though RDF makes modeling a wide range of data structures possible, it is unlikely that this flexibility will be used in the creation of complex data models for profiles. In the simplest form, each table in the CC/PP is a collection of RDF statements with simple, atomic properties. These tables may be constructed from default settings, persistent local changes, or temporary changes made by a user. One extension to the simple table of properties data model is the notion of a separate, subordinate collection of default properties. Default settings might be properties defined by the vendor. In the case of hardware, the vendor often has a very good idea of the physical properties of any given model of product. However, the current owner of the product may be able to add options, such as memory or persistent store or additional I/O devices that add new properties or change the values of some original properties. These would be persistent local changes. An example of a temporary change would be turning sound on or off.

When used in the context of a Web-browsing application, a CC/PP should be associated with a notion of a current session rather than a user or a node. HTTP and WSP both define different session semantics. The client, server, gateways, and proxies may already have their own, well-defined notions of what constitutes a connection or a session. The protocol strategy is to send as little information as possible and if anyone is missing something, they have to ask for it. If there is good reason to believe that someone is going to ask for a profile, the client can elect to send the most efficient form of the profile that makes sense.

The goal of the CC/PP framework is to specify how client devices express their capabilities and preferences (the user agent profile) to the server that originates content (the origin server). The origin server uses the user agent profile to produce and deliver content appropriate to the client device. In addition to computer-based client devices, particular attention is paid to other kinds of devices such as mobile phones.

The requirements on the framework emphasize three aspects: flexibility, extensibility, and distribution. The framework must be flexible, since we cannot today predict all the different types of devices that will be used in the future, or the ways that those devices will be used. It must be extensible for the same reasons: it should not be hard to add and test new descriptions and it must be distributed, since relying on a central registry might make it inflexible.

8

Architecture of wireless LANs

In a wireless LAN (WLAN), the connection between the client and user exists through the use of a wireless medium such as Radio Frequency (RF) or Infrared (IR) communications. This allows the mobile user to stay connected to the network. The wireless connection is most usually accomplished by the user having a handheld terminal or a laptop computer that has an RF interface card installed inside the terminal or through the PC Card slot of the laptop. The client connection from the wired LAN to the user is made through an Access Point (AP) that can support multiple users simultaneously. The AP can reside at any node on the wired network and performs as a gateway for wireless users' data to be routed onto the wired network.

The range of these systems depends on the actual usage and environment of the system but varies from 100 ft inside a solid walled building to several 1000 ft outdoors, in direct Line of Sight (LOS). This is of a similar order of magnitude as the distance that can be covered by the wired LAN in a building. However, much like a cellular telephone system, the WLAN is capable of roaming from the AP and reconnecting to the network through other APs residing at other points on the wired network. This allows the wired LAN to be extended to cover a much larger area than the existing coverage by the use of multiple APs, for example, in a university campus environment.

A WLAN can be used independently of a wired network, and it may be used as a stand-alone network anywhere to link multiple computers without having to build or extend a wired network. For example, in an outside auditing group in a client company, each auditor has a laptop equipped with a wireless client adapter. A peer-to-peer workgroup can be immediately established to transfer or access data. A member of the workgroup can be established as the server, or the network can perform in a peer-to-peer mode.

A WLAN is capable of operating at speeds in the range of 1, 2, or 11 Mbps, depending on the actual system. These speeds are supported by the standard for WLAN networks defined by the international body, the IEEE.

WLANs are billed on the basis of installed equipment cost; however, once in place there are no charges for the network use. The network communications use a part of the radio spectrum that is designated as license-free. In this band, of 2.4 to 2.5 GHz, the users can operate without a license when they use equipment that has been approved for use in this license-free band. In the United States, this license is granted by the Federal Communications Commission (FCC) for operation under Part 15 regulations. The 2.4-GHz band has been designated as license-free by the International Telecommunications Union (ITU) and is available for use, license-free in most countries in the world. The rules of operation are different in almost every country but they are similar enough so that the products can be programmed for use in every country without changing the hardware component.

The ability to build a dynamically scalable network is critical to the viability of a WLAN, as it will inevitably be used in this mode. The interference rejection of each node will be the limiting factor to the expandability of the network and its user density in a given environment.

8.1 RADIO FREQUENCY SYSTEMS

Radio Frequency (RF) and Infrared (IR) are the main technologies used for wireless communications. RF and IR technologies are used for different applications and have been designed into products that optimize the particular features of advantage.

RF is very capable of being used for applications in which communications are not line of sight and are over longer distances. The RF signals travel through walls and communicate where there is no direct path between the terminals. In order to operate in the license-free portion of the spectrum called the *Industrial, Scientific, and Medical* (ISM) *band*, the radio system must use a modulation technique called *Spread Spectrum* (SS). In this mode a radio is required to distribute the signal across the entire spectrum and cannot remain stable on a single frequency. No single user can dominate the band, and collectively all users look like noise. Spread Spectrum communications were developed to be used for secure communication links. The fact that such signals appear to be noise in the band means that they are difficult to find and to jam. This technique operates well in a real WLAN application in this band and is difficult to intercept, thus increasing security against unauthorized listeners.

The use of Spread Spectrum is especially important as it allows many more users to occupy the band at any given time and place than if they were all static on separate frequencies. With any radio system, one of the greatest limitations is available bandwidth, and so the ability to have many users operate simultaneously in a given environment is critical for the successful deployment of WLAN.

There are several bands available for use by license-free transmitters; the most commonly used are at 902 to 928 MHz, 2.4 to 2.5 GHz, and 5.7 to 5.8 GHz. Of these, the most useful is probably the 2.4-GHz band as it is available for use throughout most parts of the world. In recent years, nearly all the commercial development and the basis for the new IEEE standard has been in the 2.4-GHz band. While the 900-MHz band is widely used for other systems, it is only available in the United States and has greatly limited bandwidth available. In the license-free bands, there is a strict limit on the broadcast power of any transmitter so that the spectrum can be reused at a short distance away without interference from a distant transmitter. This is similar to the operation of a cellular telephone system.

8.2 INFRARED SYSTEMS

Another technology that is used for WLAN systems is Infrared, in which the communication is carried by light in the invisible part of the spectrum. This system is primarily of use for very short distance communications, less than 3 ft where there is a LOS connection. It is not possible for the IR light to penetrate any solid material; it is even attenuated greatly by window glass, so it is really not a useful technology in comparison to Radio Frequency for use in a WLAN system.

The application of Infrared is as a docking function and in applications in which the power available is extremely limited, such as a pager or PDA. The standard for such products is called *Infrared Data Association* (IrDA), which has been used by Hewlett Packard, IBM, and others. This is found in many notebook and laptop PCs and allows a connectionless docking facility up to 1 Mbps to a desktop machine up to two feet line of sight.

Such products are point-to-point communications and offer increased security, as only the user to whom the beam is directed can pick it up. Attempts to provide wider network capability by using a diffused IR system in which the light is distributed in all directions have been developed and marketed, but they are limited to 30 to 50 ft and cannot go through any solid material. There are very few companies pursuing this implementation. The main advantage of the point-to-point IR system – increased security – is undermined here by the distribution of the light source as it can now be received by anybody within range, not just the intended recipient.

8.3 SPREAD SPECTRUM IMPLEMENTATION

There are two methods of Spread Spectrum modulation that are used to comply with the regulations for use in the ISM band: Direct Sequence Spread Spectrum (DSSS), and Frequency Hopping Spread Spectrum (FHSS).

8.3.1 Direct sequence spread spectrum

Historically, many of the original systems available used DSSS as the required spread spectrum modulation because components and systems were available from the Direct Broadcast Satellite industry, in which DSSS is the modulation scheme used. However, the majority of commercial investments in WLAN systems are now in FHSS and the user base of FHSS products will exceed that of DSSS. Most of the new WLAN applications will be in FHSS.

A DSSS system takes a signal at a given frequency and spreads it across a band of frequencies where the center frequency is the original signal. The spreading algorithm, which is the key to the relationship of the spread range of frequencies, changes with time in a pseudorandom sequence that appears to make the spread signal a random noise source. The strength of this system is that when the ratio between the original signal bandwidth and the spread signal bandwidth is very large, the system offers great immunity to interference. For instance, if a 1-Kbps signal is spread across 1 GHz of spectrum, the spreading ratio is one million times or 60 dB. This is the type of system developed for strategic military communications systems as it is very difficult to find and is even more difficult to jam.

However, in an environment such as WLAN in the license-free, ISM band, in which the available bandwidth critically limits the ratio of spreading, the advantages that the DSSS method provides against interference become greatly limited. A realistic example in use today is a 2-Mbps data signal that is spread across 20 MHz of spectrum and that offers a spreading ratio of 10 times. This is only just enough to meet the lower limit of processing gain, a measure of this spreading ratio, as set by the FCC, the United States government body that determines the rule of operation of radio transmitters. This limitation significantly undermines the value of DSSS as a method to resist interference in real WLAN applications.

8.3.2 Frequency hopping spread spectrum

FHSS is based on the use of a signal at a given frequency that is constant for a small amount of time and then moves to a new frequency. The sequence of different channels determined for the hopping pattern, that is, where the next frequency will be to engage with this signal source, is pseudorandom. Pseudo means that a very long sequence code is used before it is repeated, over 65 000 hops, making it appear to be random. This makes it very difficult to predict the next frequency at which such a system will stop and transmit or receive data, as the system appears to be a random noise source to an unauthorized listener. This makes the FHSS system very secure against interference and interception. In an FHSS system at a data rate of 1 Mbps or higher, even a fraction of a second provides significant overall throughput for the communications system.

This system is a very robust method of communicating as it is statistically close to impossible to block all the frequencies that can be used and as there is no spreading ratio requirement that is so critical for DSSS systems. The resistance to interference is determined by the capability of the hardware filters that are used to reject signals other than the frequency of interest, and not by mathematical spreading algorithms. In the case of a standard FHSS WLAN system, with a two-stage receive section, the filtering will be provided in excess of 100 000 times rejection of unwanted signals, or over 50 dB.

8.3.3 WLAN industry standard

Industry standards are critical in the computer business and its related industries. They are the vehicles that provide a large enough market to be realistically defined and targeted with a single, compatible technological solution that many manufacturers can develop. This process reduces the cost of the products to implement the standard, which further expands the market.

In 1990, the IEEE 802 standards groups for networking set up a specific group to develop a WLAN standard similar to the Ethernet standard. In 1997, the IEEE 802.11 WLAN Standard Committee approved the IEEE 802.11 specification. This is critical for the industry as it now provides a solid specification for the vendors to target, both for systems products and components. There are three sections of the specification representing FHSS, DSSS, and IR physical layers.

The standard is a detailed software, hardware, and protocol specification with regard to the physical and data link layer of the Open System Interconnection (OSI) reference model that integrates with existing wired LAN standards for a seamless roaming environment. It is specific to the 2.4-GHz band and defines two levels of modulation that provide a basic 1-Mbps and enhanced 2-Mbps system.

The implications of an agreed standard are very significant and are the starting point for the WLAN industry in terms of a broader market. To this point, the market has been dominated by implementations that are custom developments using a specific manufacturers proprietary protocol and system. The next generation of these products for office systems will be based on the final rectified standard.

The WLAN systems discussed and those specified by the IEEE 802.11 standard operate in the unlicensed spectrum. The unlicensed spectrum allows a manufacturer to develop a piece of equipment that operates to meet predefined rules and for any user to operate the equipment without a requirement for a specific user license. This requires the manufacturer to make products that conform to the regulations for each country of operation and they should also conform to the IEEE 802.11 standard.

While the 2.4-GHz band is available in most countries, each country's regulatory bodies have usually set requirements that are different in detail. There are three major specification groups that set the trend that most other countries follow. The FCC sets a standard covered by the Part 15 regulations that are used in much of the rest of the United States and the world. The Japanese Nippon Telegraph and Telephone (NTT) has its own standard. The European countries have set a specification through European Telecommunications Standards Institute (ETSI).

While all these differ in detail, it is possible to make a single hardware product that is capable of meeting all three specifications with only changes to the operating software. Although the software could be downloaded from a host such as a notebook PC, the changes are required to be set by the manufacturer and not the user in order to meet the rules of operation.

The increasing demand for network access while mobile will continue to drive the demand for WLAN systems. The Frequency Hopping technology has the ability to support significant user density successfully, so there is no limitation to the penetration of such products in the user community. WLAN solutions will be especially viable in new markets such as the Small Office/Home Office (SOHO) market, where there is rarely a wired LAN owing to the complexity and cost of wiring. WLAN offers a solution that will connect a generation to wired access, but without using the wires.

8.4 IEEE 802.11 WLAN ARCHITECTURE

In IEEE 802.11 the addressable unit is a station (STA), which is a message destination, but not (in general) a fixed location. IEEE 802.11 handles both mobile and portable stations. Mobile Stations (MSs) access the LAN while in motion, whereas a Portable Station (PS) can be moved between locations, but it is used only at a fixed location. MSs are often battery powered, and power management is an important consideration since we cannot assume that a station's receiver will always be powered on.

IEEE 802.11 appears to higher layers [logical link control (LLC)] as an IEEE 802 LAN, which requires that the IEEE 802.11 network handles station mobility within the Medium Access Control (MAC) sublayer, and meets the reliability assumptions that LLC makes about the lower layers.

The IEEE 802.11 architecture provides a WLAN supporting station mobility transparently to upper layers. The Basic Service Set (BSS) is the basic building block consisting of member stations remaining in communication. If a station moves out of its BSS, it can no longer directly communicate with other members of the BSS.

The Independent BSS (IBSS) is the most basic type of IEEE 802.11 LAN and may consist of at least two stations that can communicate directly. This LAN is formed only as long as it is needed and is often referred to as an *ad hoc network*. The association between an STA and a BSS is dynamic, since STAs turn on and off, come within range and go out of range.

A BSS may form the Distribution System (DS), which is an architectural component used to interconnect BSSs. IEEE 802.11 logically separates the Wireless Medium (WM) from the Distribution System Medium (DSM). Each logical medium is used for different purposes by a different component of the architecture. The IEEE 802.11 LAN architecture is specified independently of the physical characteristics of any specific implementation. The DS enables mobile device support by providing the logical services necessary to handle address-to-destination mapping and seamless integration of multiple BSSs. An Access Point (AP) is an STA that provides access to the DS by providing DS services in addition to acting as an STA. The data move between a BSS and the DS *via* an AP. All APs are also STAs, and they are addressable entities. The addresses used by an AP for communication on the WM and on the DSM are not necessarily the same.

The DS and BSSs allow IEEE 802.11 to create a wireless network of arbitrary size and complexity called *Extended Service Set* (ESS) *network*. The ESS network appears the same to an LLC layer as an IBSS network. Stations within an ESS may communicate and MSs may move from one BSS to another (within the same ESS), transparently to LLC.

A portal is the logical point at which MAC Service Data Units (MSDUs) from an integrated non-IEEE 802.11 LAN enter the IEEE 802.11 DS. All data from non-IEEE 802.11 LANs enter the IEEE 802.11 architecture *via* a portal, which provides logical integration between the IEEE 802.11 architecture and existing wired LANs. A device may offer both the functions of an AP and a portal, for example, when a DS is implemented from IEEE 802 LAN components.

Architectural services of IEEE 802.11 are as follows: authentication, association, deauthentication, disassociation, distribution, integration, privacy, reassociation, and MSDU delivery. These services are provided either by stations as the Station Service (SS) or by the DS as the Distribution System Service (DSS).

The SS includes authentication, deauthentication, privacy, and MSDU delivery. The SS is present in every IEEE 802.11 station, including APs, and is specified for use by MAC sublayer entities.

The DSSs include association, disassociation, distribution, integration, and reassociation. The DSSs are provided by the DS and are accessed by an AP, which is an STA that also provides DSSs. DSSs are specified for use by MAC sublayer entities.

The IEEE 802.11 architecture handles multiple logical media and address spaces and is independent of the DS implementation. This architecture interfaces cleanly with network layer mobility approaches.

8.4.1 IEEE 802.11a and IEEE 802.11b

IEEE 802.11a is an extension to 802.11 that applies to WLANs and provides up to 54 Mbps in the 5-GHz band. IEEE 802.11a uses an Orthogonal Frequency Division Multiplexing (OFDM) encoding scheme rather than FHSS or DSSS.

The IEEE 802.11a standard is designed to operate in the 5-GHz Unlicensed National Information Infrastructure (UNII) band. Specifically, the FCC has allocated 300 MHz of spectrum for unlicensed operation in the 5-GHz block, 200 MHz of which is at 5.15 to 5.35 MHz, with the other 100 MHz at 5.725 to 5.825 MHz. The spectrum is split into three working domains. The first 100 MHz in the lower section is restricted to a maximum power output of 50 mW (milliwatts). The second 100 MHz has 250-mW power output, and the top 100 MHz is used for outdoor applications with a maximum of 1 watt power output.

IEEE 802.11b, also referred to as 802.11 High Rate or Wi-Fi (Wireless Fidelity), is an extension to 802.11 that applies to WLANs and provides 11-Mbps transmission (with a fallback to 5.5, 2, and 1 Mbps) in the 2.4-GHz band. IEEE 802.11b uses only DSSS. IEEE 802.11b was a 1999 ratification to the original 802.11 standard, allowing wireless functionality comparable to Ethernet.

The IEEE 802.11b specification allows for the wireless transmission of approximately 11 Mbps of data at distances from several dozen to several 100 ft over the 2.4-GHz (2.4 to 2.483) unlicensed RF band. The distance depends on impediments, materials, and LOS.

IEEE 802.11b standard defines two bottom levels of OSI reference model – the Physical Layer (PHY) and the Data Link Layer (MAC sublayer).

IEEE 802.11b defines two pieces of equipment, a wireless station, which is usually a PC or a Laptop with a wireless Network Interface Card (NIC), and an Access Point (AP), which acts as a bridge between the wireless stations and Distribution System or wired networks. There are two operation modes in IEEE 802.11b, Infrastructure Mode and Ad Hoc Mode.

Infrastructure Mode consists of at least one AP connected to the Distribution System. An AP provides a local bridge function for the BSS. All wireless stations communicate with the AP and no longer communicate directly. All frames are relayed between wireless stations by the AP.

An ESS is a set of infrastructure BSSs, in which the APs communicate amongst themselves to forward traffic from one BSS to another to facilitate movement of wireless stations between BSSs.

The wireless stations communicate directly with each other. Every station may not be able to communicate with every other station because of the range limitations. There are no APs in an IBSS. Therefore all stations need to be within range of each other and they communicate directly.

IEEE 802.11b defines dynamic rate shifting, allowing data rates to be automatically adjusted for noisy conditions. This means IEEE 802.11b devices will transmit at lower speeds, 5.5 Mbps, 2 Mbps, and 1 Mps under noisy conditions. When the devices move back within range of a higher speed transmission, the connection will automatically speed up again.

8.5 BLUETOOTH

Bluetooth devices operate at 2.4 GHz in the ISM band. The operating band of 83.5 MHz is divided into 1-MHz channels, each signaling data at 1 M Symbols per second to obtain 1 Mb s^{-1} available channel bandwidth by using Gaussian Frequency Shift Keying (GFSK). Bluetooth devices use FHSS, and each time slot lasts 625 μs. The radio power ranges for Bluetooth applications are 10 m, 20 m, and 100 m, for different power classes.

Bluetooth devices operate in two modes, as a master or as a slave. The master sets the frequency hopping sequence, and the slaves synchronize to the master in time and frequency. Every Bluetooth device has a unique address and clock. All slaves use the master address and clock and are synchronized to the master's frequency hop sequence. Each Bluetooth device may be either a master or a slave at any time, but not simultaneously. A master initiates an exchange of data and a slave responds to the master.

The master controls when devices are allowed to transmit. The master allows slaves to transmit by allocating slots for data traffic or voice traffic. In data traffic slots, the slaves are only allowed to transmit when replying to a transmission to them by the master. In voice traffic slots, slaves are required to transmit regularly in reserved slots whether or not they are replying to the master. The master uses Time Division Multiplexing (TDM) to allocate time slots to the slaves depending on the data-transfer requirements.

A number of slave devices operating together with one master create a piconet in which all devices follow the frequency hopping sequence and timing of the master. The slaves in a piconet only have links to the master; there are no direct links between slaves. A point-to-point connection occurs with one slave and a master, and a point-to-multipoint connection exists with one master and up to seven slaves in a piconet.

Piconets can be linked into a scatternet, in which some devices are members of more than one piconet. These devices must time-share, spending a few slots on each piconet. Different devices are the masters in different piconets, and a scatternet cannot share a master.

Bluetooth devices sharing a piconet are synchronized to avoid collision, but the devices in other piconets are not synchronized and may randomly collide on the same frequency. The packets lost because of collision will be retransmitted and voice packets will be ignored. The number of collisions and retransmissions increases with the growing number of piconets and scatternets.

Three power classes allow Bluetooth devices to connect at different ranges: Class 1 uses 100 mW (20 dBm), Class 2 uses 2.5 mW (4 dBm), Class 3 uses 1 mW (0 dBm). The maximum ranges for Classes 1, 2, and 3, are 100 m, 20 m, and 10 m, respectively.

Bluetooth devices use Asynchronous Connectionless (ACL) links for data communication and Synchronous Connection Oriented (SCO) links for voice communication.

The ACL link provides a packet-switched connection in which data is exchanged sporadically as and when data is available. The master decides, on a slot-by-slot basis, as to which slave to transmit, or from which slave to receive information. This way both asynchronous and isochronous (time-bounded) services are possible. Most ACL packets facilitate error checking and retransmission to assure data integrity. If a slave is addressed in a master to slave slot, this slave may only respond with an ACL packet in the next slave to master slot. If the slave fails to decode the slave address in the packet header, or it is not sure whether it was addressed, then it is not allowed to respond. Broadcast packets are ACL packets not addressed to a specific slave and are received by every slave.

The SCO link provides a circuit-switched connection between master and slave. A master can support up to three SCO links to the same slave or to different slaves. A slave can support up to three SCO links from the same master. Because of the delay-bounded nature of SCO data, SCO packets are never retransmitted. The master will transmit SCO packets to the slave at regular intervals, counted in slots. The slave is allowed to respond with an SCO packet in the reserved response slot, unless it correctly decoded the packet header and discovered that it had not been addressed as expected. If the packet was incorrectly decoded because of errors, then the slave may still respond as the slot is reserved and the master is not allowed to transmit elsewhere in the reserved slot. An exception is a broadcast Link Management Protocol (LMP) message, which takes precedence over the SCO link.

A Bluetooth packet consists of an access code, a header, and a payload. The access code is used to detect the presence of a packet and to address the packet to a specific device. For example, slaves detect the presence of a packet by matching the access code against their stored copy of the master's access code. The header contains the control information associated with the packet and the link, such as the address of the slave to which the packet is sent. The payload contains the actual message information, if this is a higher layer protocol message, or the data, if this is data being passed down the stack.

The following four access codes are used by Bluetooth:

- Channel Access Code (CAC) derived from the master's Lower Address Part (LAP) and used by all devices in a piconet during the exchange of data over a live connection.
- Device Access Code (DAC) derived from a specific device's LAP. DAC is used when paging a specific device and by that device in Page Scan while listening for paging messages to itself.
- General Inquiry Access Code (GIAC) used by all devices during the inquiry procedures, since no prior knowledge of anyone's LAP exists.
- Dedicated Inquiry Access Code (DIAC) is a specified range of inquiry access code (IAC) reserved to carry inquiry procedures between specific sets of devices like printers or cellular handsets.

8.5.1 Bluetooth architecture

The Bluetooth protocol stack is shown in Figure 8.1 opposite the Open Systems Interconnect (OSI) standard reference model. The physical layer covers the radio and part of the baseband and is responsible for the electrical interface to the communications media, including modulation and channel coding. The data link layer covers the control end of the baseband, including error checking and corrections and overlaps the link controller

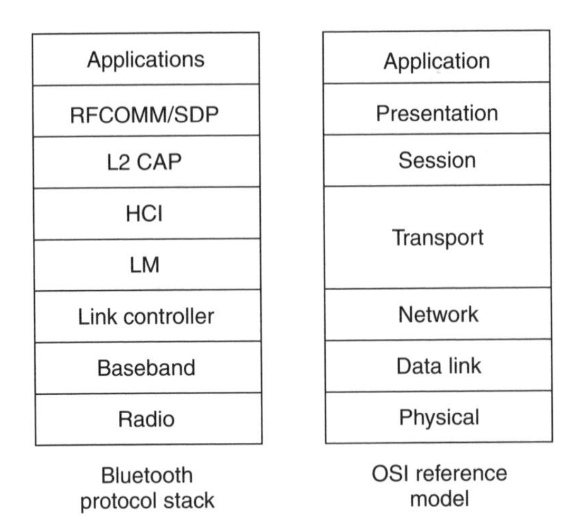

Figure 8.1 The Bluetooth protocol stack and OSI reference model.

function. The data link layer is responsible for transmission, framing, and error control over a particular link.

The network layer covers the higher end of the link controller, setting up and maintaining multiple links, and most of the Link Manager (LM) functions. The network layer is responsible for data transfer across the network, independent of the media and specific topology of the network. The transport layer covers the high end of the LM and overlaps the Host Controller Interface (HCI). The transport layer is responsible for reliability and multiplexing of data transfer across the network to the level provided by the application.

The session layer covers Logical Link Control and Adaptation Protocol (L2CAP) and the lower end of RFCOMM/SDP, where RFCOMM is a protocol for RS-232 serial cable emulation and Service Discovery Protocol (SDP) is a Bluetooth protocol that allows a client to discover the devices offered by a server. The session layer provides the management and data flow control services.

The presentation layer covers the main functions of RFCOMM/SDP and provides a common representation for the application layer data by adding service structure to the units of data. The application layer is responsible for managing communications between host applications.

A Bluetooth device is in one of the following states:

- Standby state, in which the device is inactive, no data is being transferred, and the radio is not switched on. In this state the device is unable to detect any access codes.
- Inquiry state, when a device attempts to discover all the Bluetooth-enabled devices in its local area.
- Inquiry scan is used by devices to make themselves available to inquiring devices.
- Page state is entered by the master, which transmits paging messages to the intended slave device.
- Page scan is used by a device to allow paging devices to establish connection with it.

- Connection–Active is a state in which the slave moves to the master's frequency hop and timing sequence by switching to the master's clock.
- Connection–Hold mode allows the device to maintain its active member address while ceasing to support ACL traffic for a certain time to free bandwidth for other operations such as scanning, paging, inquiry, or low-power sleep.
- Connection–Sniff mode allows the device to listen for traffic by using a predefined slot time.
- Connection–Park mode allows the device to enter low-power sleep mode by giving up its active member address and listening for traffic only occasionally.

The SDP is used for device discovery. The messages exchanged between two devices during inquiry and inquiry scan are shown in Figure 8.2. The inquiring device calls out by transmitting Identifier (ID) packets containing an IAC. When a scanning device first hears an inquiry, it waits for a random period and then reenters the scanning state, listening once more for another ID. This time, if it hears the inquiry, it replies with the Frequency Hop Synchronization (FHS) packet. Several devices can respond to an inquiry but their responses are spaced out randomly and do not interfere. The inquirer must keep inquiring for longer than the range of the random period.

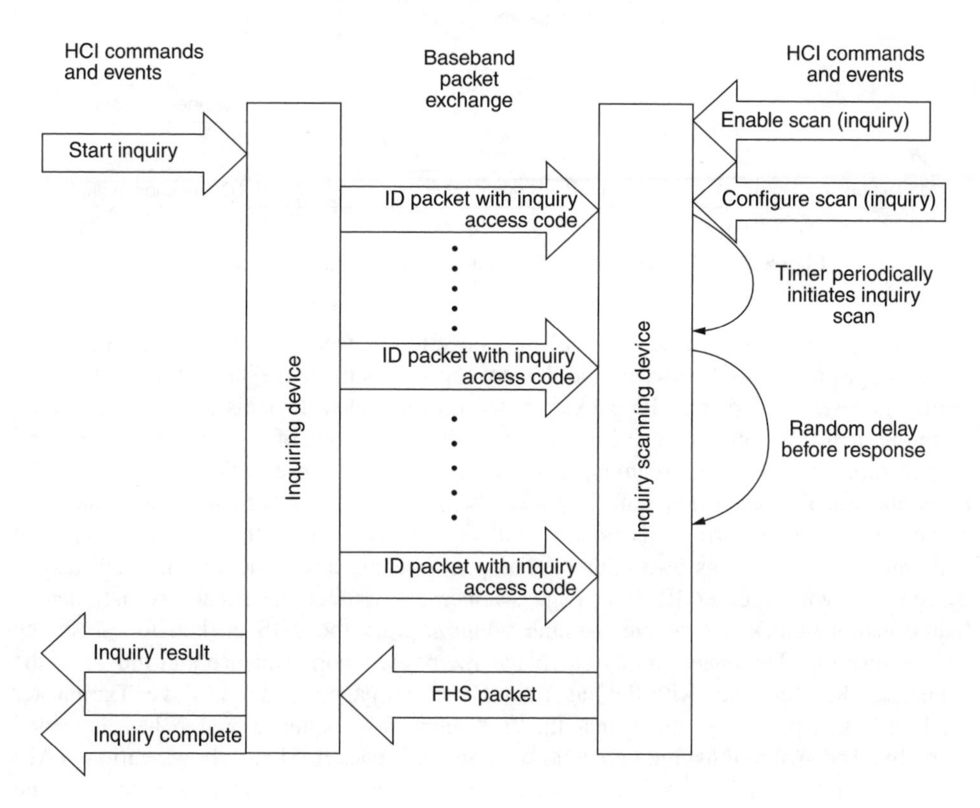

Figure 8.2 Message sequence chart for inquiry and inquiry scan.

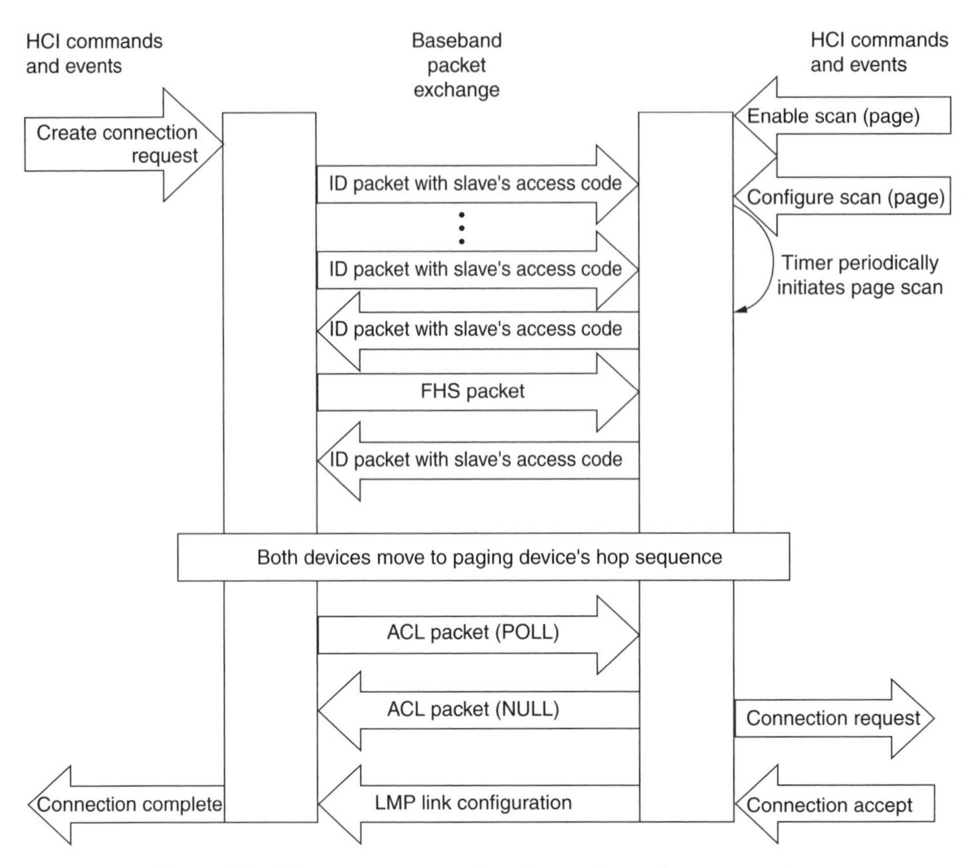

Figure 8.3 Message sequence chart for paging and page scanning.

A paging procedure is used to establish connection between devices. Figure 8.3 shows the messages exchanged between two devices during connection establishment. A device sends out a series of paging ID packets based on the paged device's address. The page-scanning device is configured to carry out periodic page scans of a specified duration and at a specified interval. The scanning device starts a timer and a periodic scan when the timer elapses. The pager transmits ID packets with the page scanner's address. If the page scanner is scanning during this time, it will trigger and receive the ID packet replying with another ID, using its own address. The page scanner acknowledges the FHS packet by replying with another ID. The page scanner can extract the necessary parameters, like Bluetooth clock and active member address, from the FHS packet, to use in the new connection. The page scanner calculates the pager's hop sequence and moves to the connection hop sequence with the pager as master and page scanner as slave. The master sends a POLL packet to check that the frequency hop sequence switch has happened correctly. The switch must then respond with an ACL packet. Then follows various LMP link configuration packets exchange. If required, master and slave can agree to swap the roles in a master/slave switch.

A connection between a host and a Bluetooth device is established in the following steps:

1. Host requests an inquiry.
2. Inquiry is sent using the inquiry hopping sequence.
3. Inquiry scanning devices respond to the inquiry scan with FHS packets that contain all the information needed to connect with them.
4. The contents of the FHS packets are passed back to the host.
5. The host requests connection to one of the devices that responded to the inquiry.
6. Paging is used to initiate a connection with the selected device.
7. If the selected device is page scanning, it responds to the page.
8. If the page-scanning device accepts the connection, it will start hopping using the master's frequency hopping sequence and timing.

The following operations are managed by the link manager (LM), which translates the Host Controller Interface (HCI) commands.

1. Attaching slaves to a piconet and allocating their active member addresses
2. Breaking connections to detach slaves from a piconet
3. Configuring the link including controlling master/slave switches in which both devices must simultaneously change roles
4. Establishing ACL data and SCO voice links
5. Putting connections into low power modes: hold, sniff, and park
6. Controlling test modes.

In Bluetooth standard the HCI uses command packets for the host to control the module. The module uses event packets to inform the host of changes in the lower layers. The data packets are used to pass voice and data between host and module. The transport layers carry HCI packets.

The host controls a Bluetooth module by using the following HCI commands:

1. Link control, for instance, setting up, tearing down, and configuring links.
2. Power-saving modes and switching of the roles between master and slave.
3. Direct access to information on the local Bluetooth module and access to information on remote devices by triggering LMP exchanges.
4. Control of baseband features, for instance, timeouts.
5. Retrieving status information on a module.
6. Invoking Bluetooth test modules for factory testing and for Bluetooth qualification.

L2CAP sends data packets to HCI or to LM. The functions of L2CAP include

- multiplexing between higher layer protocols that allows them to share lower layer links;
- segmentation and reassembly to allow transfer of larger packets than those that lower layers support;
- group management by providing one-way transmission to a group of other Bluetooth devices;
- quality of service management for higher layer protocols.

L2CAP provides the following facilities needed by higher layer protocols:

- Establishing links across underlying ACL channels using L2CAP signals;
- Multiplexing between different higher layer entities by assigning each one its own connection ID;
- Providing segmentation and reassembly facilities to allow large packets to be sent across Bluetooth connections.

RFCOMM is a protocol for RS-232 serial cable emulation. It is a reliable transport protocol with framing, multiplexing, and providing modem status, remote line status, remote port settings, and parameter negotiation. RFCOMM supports devices with internal emulated serial port and intermediate devices with physical serial port. RFCOMM communicates with frames that are carried in the data payload in L2CAP packets. An L2CAP connection must be set up before an RFCOMM connection can be set up. RFCOMM is based on the Global System for Mobile communications (GSM) 07.10 standard with a few minor differences between Bluetooth and GSM cellular phone connections.

SDP server is a Bluetooth device offering services to other Bluetooth devices. Each SDP server maintains its own database containing information about those services. An L2CAP channel must be established between SDP client and server, which uses a protocol service multiplexor reserved for SDP. After the SDP information has been retrieved from the server, the client must establish a separate connection to use the service. Services have Universally Unique Identifiers (UUIDs) that describe them.

8.5.2 Bluetooth applications

Bluetooth can be used as a bearer layer in WAP architecture. A WAP server can be preconfigured with the Bluetooth device address of a WAP server in range, or a WAP client can find it by conducting an inquiry and then use service discovery to find a WAP server. The WAP client needs to use SDP to find the RFCOMM server number allocated to WAP services. The WAP server's service discovery record identifies whether the service is a proxy used to access files on other devices, an origin server, which provides its own files, or both. The provided information includes home URL, service name, and a set of parameters needed to connect to the WAP service, which are the port numbers allocated to the various layers of the WAP stack. Once the service discovery information has been retrieved, an L2CAP link for RFCOMM is established and a WSP session is set up over this link. The URLs can be requested by Wireless Application Environment (WAE) across WSP.

The WAP layers use User Datagram Protocol (UDP), Internet Protocol (IP), and Point-to-Point Protocol (PPP), which allow datagrams to be sent across Bluetooth's RFCOMM serial port emulation layer. The WAP components used above the Bluetooth protocol stack are shown in Figure 8.4.

Telephony Control protocol Specification (TCS) provides call control signaling to establish voice and data calls between Bluetooth devices. TCS signals use L2CAP channel with a Protocol and Service Multiplexor (PSM) reserved for TCS. A separate bearer channel is established to carry the call, for example, an SCO channel or an ACL channel. TCS

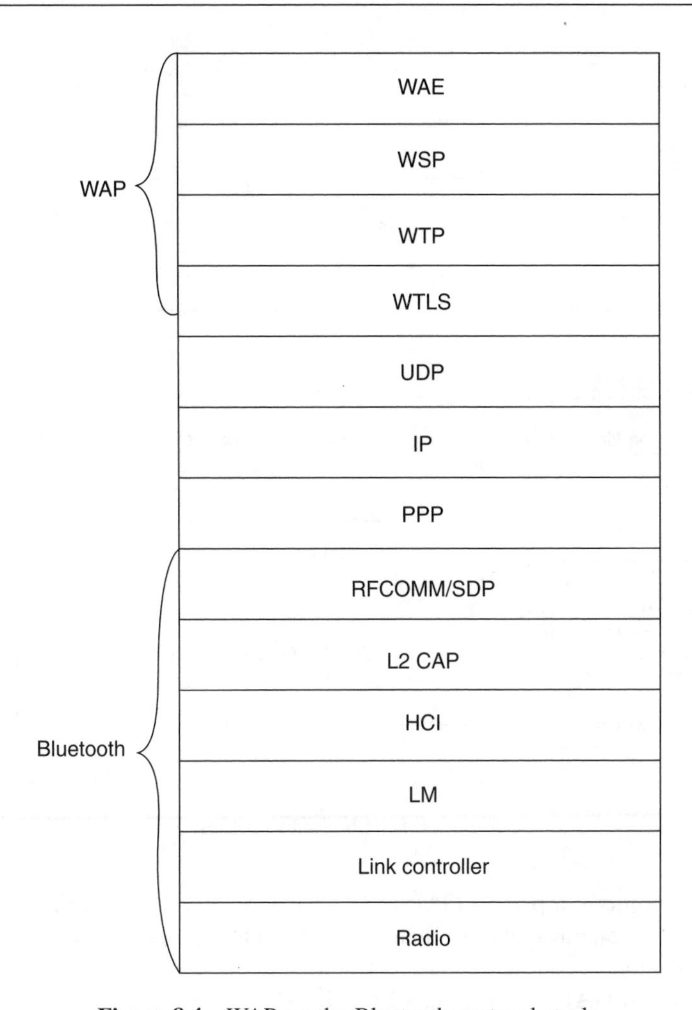

Figure 8.4 WAP on the Bluetooth protocol stack.

does not define handover of calls from one device to another and does not provide a mechanism for groups of devices to enter into conference calls; only point-to-point links are supported.

Bluetooth profiles are illustrated in Figure 8.5. The Generic Access Profile (GAP) is the basic Bluetooth profile used by the devices to establish baseband links.

The serial port profile provides RS-232 serial cable emulation for Bluetooth devices. It provides a simple, standard way for applications to interoperate, for example, legacy applications do not need to be modified to use Bluetooth; they can treat Bluetooth link as a serial cable link. The serial port profile is based on the GSM standard GSM 07.10, which allows multiplexing of serial connections over one serial link. It supports a communication end point, for instance, a laptop. It also supports intermediate devices, which form part of a communications link, for instance, modems.

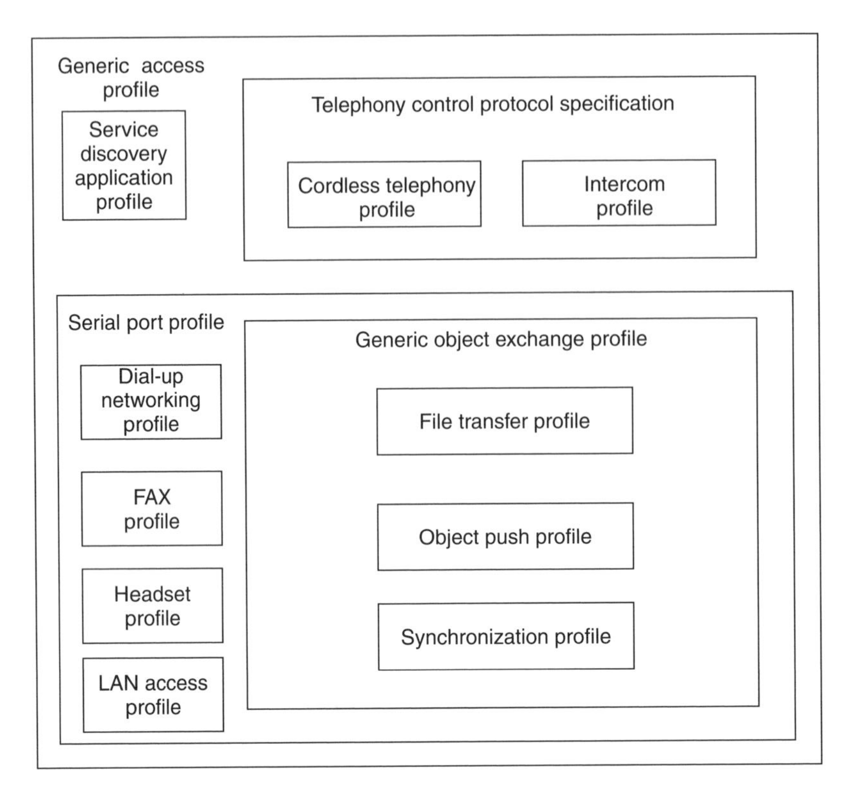

Figure 8.5 Bluetooth profiles.

Serial port profile is a part of GAP and consists of dial-up networking, fax, headset, LAN access, and general object exchange profile, which uses file transfer, object push, and synchronization.

TCS is a part of GAP and consists of cordless telephony and intercom.

The cordless telephony profile defines a gateway that connects to an external network and receives incoming calls, and a terminal that receives the calls from the gateway and provides speech and/or data links to the user. The gateway is the master of a piconet connecting up to seven terminals at a time; however, because of the limitations on SCO capacity, only three active voice links can be supported simultaneously. The gateway can pass calls to the terminals or the terminals can initiate calls and route them through the gateway. This allows Bluetooth devices that do not have telephony links to access telephone networks through the gateway.

The intercom profile supports direct point-to-point voice connections between Bluetooth devices.

8.5.3 Bluetooth devices

Bluetooth devices use low power modes to keep connections, but switch off receivers to save battery power. The low power modes include hold, which allows devices to be inactive for a single short period; sniff that allows devices to be inactive except for periodic sniff slots; and park, which is similar to sniff, except that parked devices give up their active member address.

Hold mode is used to stop ACL traffic for a specified period of time, but SCO traffic is not affected. Master and slave can force or request hold mode. A connection enters hold mode due to a request from the local host, or when a link manager at the remote end of a connection requests it to hold, or when the local link manager autonomously decides to put the connection in hold mode. A device enters hold mode when all its connections are in hold mode.

Sniff mode is used to save power on low data rate links and reduces traffic to periodic sniff slots. A device in sniff mode only wakes up periodically in prearranged sniff slots. The master and slave must negotiate the timing of the first sniff slot and the interval at which further sniff slots follow. They also negotiate the window in which the sniffing slave will listen for transmissions and the sniff timeout. A device enters sniff mode due to a request from its own host, or when a link manager at the remote end of a connection requests or forces it to enter sniff mode. A master can force a slave into sniff mode and give permission to a slave, which requests to enter sniff mode.

A device entering park mode gives up its active member address and ceases to be an active member of the piconet. This device cannot transmit and cannot be addressed directly by the master; however, it wakes up periodically and listens for broadcasts, which can be used to unpark it. At prearranged beacon instants, a device in park mode wakes periodically to listen for transmissions from the master during a series of beacon slots. Park mode allows the greatest power saving.

Quality of service (QoS) capabilities include data rate, delay, and reliability and are provided by the link manager, which chooses packet types, sets polling intervals, allocates buffers, link bandwidth, and makes decisions about performing scans. Link managers negotiate peer to peer to ensure that QoS is coordinated at both ends of a link. For unicast (point-to-point), a reliable link is provided by the receiver acknowledging packets. The packets not acknowledged are retransmitted. Broadcast packets are not acknowledged, and to provide a reliable link, a Bluetooth device can be set to retransmit broadcast packets a certain number of times.

Figure 8.6 shows Bluetooth devices with different capabilities. These devices are personal devices, point-to-point devices, point-to-multipoint devices, and scatternet devices.

A personal device connects only to one other preset device. If another device attempts to connect to a personal device, it refuses the LMP connection request message.

A point-to-point device supports only one ACL data link at a time. When a Bluetooth device inquires for other devices in the area, it receives FHS packets, which contain all the information required to connect to the devices, including the clock offset. A point-to-point device must keep a database of the devices it has discovered.

Point-to-multipoint device can establish links to several devices. This device must handle QoS balancing between the links through allocation of bandwidth to each link. Bandwidth requirements are received from the higher layer protocols and lower layers send QoS violations to the higher layers.

A scatternet is a group of linked piconets joined by common members that can either be slaves on both piconets or a master of one piconet and a slave on another. Bandwidth

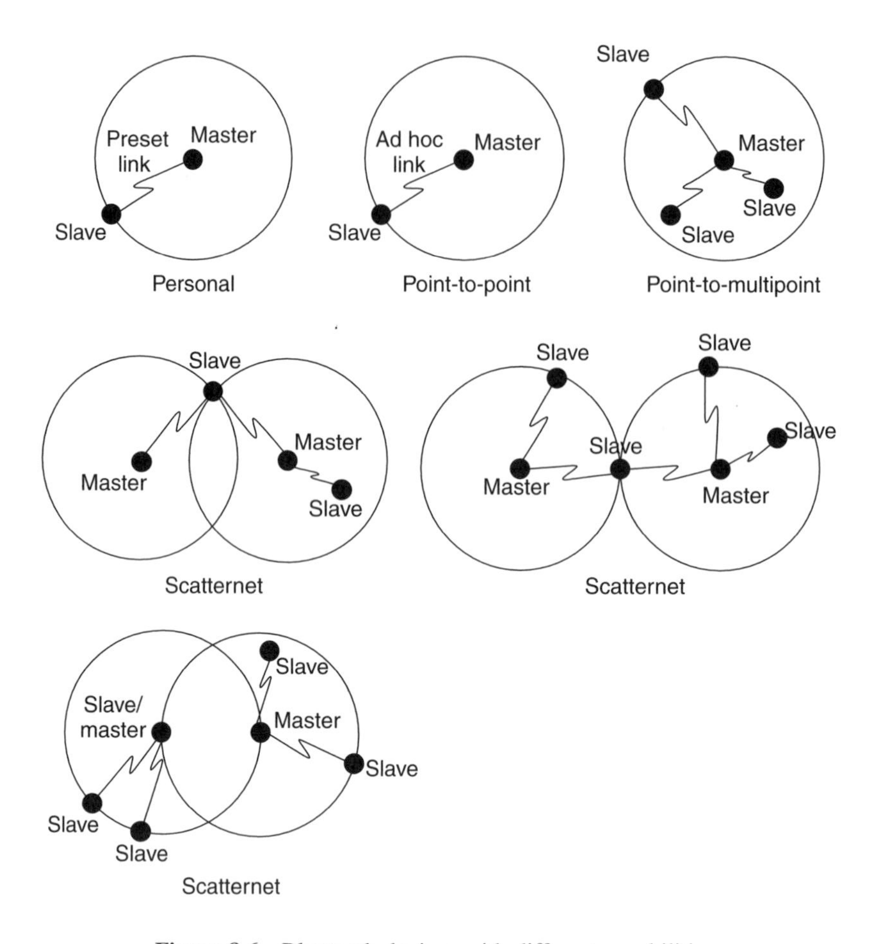

Figure 8.6 Bluetooth devices with different capabilities.

is reduced in scatternet by the time taken to switch between piconets, which should be done infrequently. Devices are absent from one piconet when present on the other, and this absence should be managed if maximum efficiency is required. Managing piconets involves calculation and negotiation of parameters for QoS, and possibly beacon slots, sniff slots, or hold times.

A device manager performs as an interpreter between applications and Bluetooth protocols as shown in Figure 8.7. Applications requesting links and discovering devices use the device manager, which also has the information about configuration control. A device manager handles set up, tear down, and configuration of the baseband links, QoS parameters and trade-offs, management of higher layer links, device discovery, and information caching. The device manager has an interface to HCI, RFCOMM, and LC2CAP.

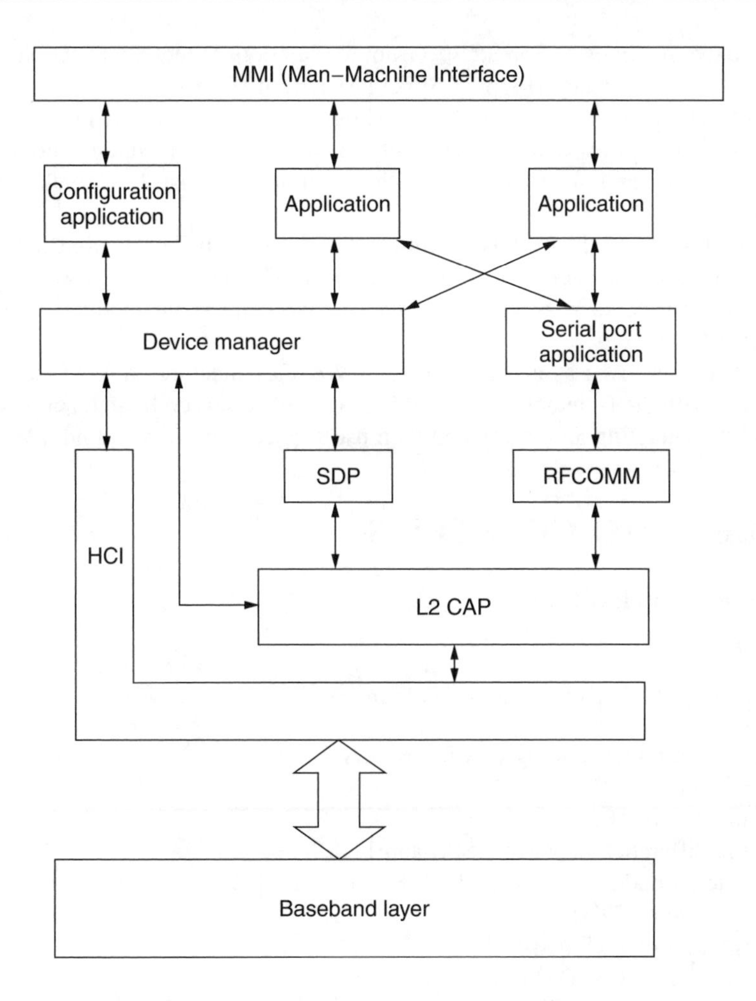

Figure 8.7 Bluetooth protocol stack with device manager.

8.6 SUMMARY

The use of Spread Spectrum is especially important as it allows many more users to occupy the band at any given time and place than if they were all static on separate frequencies. With any radio system, one of the greatest limitations is available bandwidth, and so the ability to have many users operate simultaneously in a given environment is critical for the successful deployment of WLAN.

The application of Infrared is as a docking function and in applications in which the power available is extremely limited, such as a pager or PDA.

There are two methods of Spread Spectrum modulation: Direct Sequence Spread Spectrum (DSSS) and Frequency Hopping Spread Spectrum (FHSS).

IEEE 802.11 appears to higher layers [Logical Link Control (LLC)] as an IEEE 802 LAN, which requires that the IEEE 802.11 network handle station mobility within the MAC sublayer and meets the reliability assumptions that LLC makes about the lower layers.

In Bluetooth standard the HCI uses command packets for the host to control the module. The module uses event packets to inform the host of changes in the lower layers. The data packets are used to pass voice and data between host and module. The transport layers carry HCI packets.

Bluetooth can be used as a bearer layer in WAP architecture. A WAP server can be preconfigured with the Bluetooth device address of a WAP server in range, or a WAP client can find it by conducting an inquiry and then use service discovery to find a WAP server.

PROBLEMS TO CHAPTER 8

Architecture of wireless LANs

Learning objectives

After completing this chapter, you are able to

- demonstrate an understanding of wireless LANs;
- explain the role of RF;
- explain the role of IR;
- explain the difference between DSSS and FHSS;
- demonstrate an understanding of IEEE 802.11 WLAN architecture;
- explain the role of STA;
- explain the role of BSS and ESS;
- explain the role of DS, DSM, and DSS;
- demonstrate an understanding of Bluetooth architecture;
- explain the states of a Bluetooth device;
- explain how a connection is established between a host and Bluetooth device;
- explain the role of the HCI;
- explain the functions of L2CAP.

Practice problems

8.1: What are the WLAN's operating speeds?
8.2: What is the radio frequency (RF) band in which the LANs operate?
8.3: Where is RF used?
8.4: Where is IR used?
8.5: What is the function of DSSS?
8.6: What is the function of FHSS?
8.7: What is the addressable unit in IEEE 802.11?

8.8: What is BSS?
8.9: What is the architecture of the DS?
8.10: What is the role of a portal?
8.11: What are the architectural services of IEEE 802.11?
8.12: What are the possible states of a Bluetooth device?
8.13: How is a connection between a host and Bluetooth device established?
8.14: How does the host control a Bluetooth module?
8.15: What are the functions of L2CAP?

Practice problem solutions

8.1: A WLAN is capable of operating at speeds in the range of 1, 2, or 11 Mbps depending on the actual system. These speeds are supported by the standard for WLAN networks defined by the international body, the IEEE.
8.2: WLAN communications take place in a part of the radio spectrum that is designated as license-free. In this band, 2.4 to 2.5 GHz, users can operate without a license as long as they use equipment that has been type-approved for use in the license-free bands.
8.3: RF is very capable of being used for applications in which communications are not line-of-sight and are over longer distances. The RF signals travel through walls and communicate where there is no direct path between the terminals.
8.4: Infrared is primarily used for very short distance communications, less than 3 ft where there is an LOS connection. It is not possible for the Infrared light to penetrate any solid material; it is even attenuated greatly by window glass, so it is really not a useful technology in comparison to Radio Frequency for use in a WLAN system. The application of Infrared is as a docking function and in applications in which the power available is extremely limited such as a pager or PDA. The standard for such products is called *Infrared Data Association* (IrDA), which has been used by Hewlett Packard, IBM, and others. This is found in many notebook and laptop PCs and allows a connectionless docking facility up to 1 Mbps to a desktop machine and up to two feet, line of sight.
8.5: A DSSS system takes a signal at a given frequency and spreads it across a band of frequencies where the center frequency is the original signal. The spreading algorithm, which is the key to the relationship of the spread range of frequencies, changes with time in a pseudorandom sequence that appears to make the spread signal a random noise source.
8.6: Frequency Hopping Spread Spectrum (FHSS) is based on the use of a signal at a given frequency that is constant for a small amount of time and then moves to a new frequency. The sequence of different channels determined for the hopping pattern, that is, where the next frequency will be to engage with this signal source, is pseudorandom.
8.7: In IEEE 802.11 the addressable unit is a station (STA), which is a message destination, but not (in general) a fixed location. IEEE 802.11 handles both mobile and portable stations. MSs access the LAN while in motion, whereas a Portable Station (PS) can be moved between locations but it is used only at a fixed location. MSs

are often battery powered, and power management is an important consideration since we cannot assume that a station's receiver will always be powered on.

8.8: The IEEE 802.11 architecture provides a WLAN supporting station mobility transparently to upper layers. The Basic Service Set (BSS) is the basic building block consisting of member stations remaining in communication. If a station moves out of its BSS, it can no longer directly communicate with other members of the BSS.

The IBSS is the most basic type of IEEE 802.11 LAN, and may consist of at least two stations that can communicate directly. This LAN is formed only as long as it is needed, and is often referred to as an *ad hoc* network. The association between an STA and a BSS is dynamic since STAs turn on and off, come within range and go out of range.

8.9: A BSS may form the Distribution System (DS), which is an architectural component used to interconnect BSSs. IEEE 802.11 logically separates the Wireless Medium (WM) from the Distribution System Medium (DSM). Each logical medium is used for different purposes by a different component of the architecture. The IEEE 802.11 LAN architecture is specified independently of the physical characteristics of any specific implementation. The DS enables mobile device support by providing the logical services necessary to handle address-to-destination mapping and seamless integration of multiple BSSs. An Access Point (AP) is an STA that provides access to the DS by providing DS services in addition to acting as an STA. The data move between a BSS and the DS *via* an AP. All APs are also STAs and they are addressable entities. The addresses used by an AP for communication on the WM and on the DSM are not necessarily the same.

The DS and BSSs allow IEEE 802.11 to create a wireless network of arbitrary size and complexity called *the Extended Service Set* (ESS) *network*. The ESS network appears the same to an LLC layer as an IBSS network. Stations within an ESS may communicate and MSs may move from one BSS to another (within the same ESS) transparently to LLC.

8.10: A portal is the logical point at which MAC Service Data Units (MSDUs) from an integrated non-IEEE 802.11 LAN enter the IEEE 802.11 DS. All data from non-IEEE 802.11 LANs enter the IEEE 802.11 architecture *via* a portal, which provides logical integration between the IEEE 802.11 architecture and existing wired LANs. A device may offer both the functions of an AP and a portal, for example, when a DS is implemented from IEEE 802 LAN components.

8.11: Architectural services of IEEE 802.11 are as follows: authentication, association, deauthentication, disassociation, distribution, integration, privacy, reassociation, and MSDU delivery. These services are provided either by stations as the Station Service (SS) or by the DS as the Distribution System Service (DSS).

The SS includes authentication, deauthentication, privacy, and MSDU delivery. The SS is present in every IEEE 802.11 station, including APs and is specified for use by MAC sublayer entities.

The DSSs include association, disassociation, distribution, integration, and reassociation. The DSSs are provided by the DS and accessed by an AP, which is an STA that also provides DSSs. DSSs are specified for use by MAC sublayer entities.

8.12: A Bluetooth device is in one of the following states:

- Standby state, in which the device is inactive, no data is being transferred, and the radio is not switched on. In this state the device is unable to detect any access codes.
- Inquiry state, when a device attempts to discover all the Bluetooth-enabled devices in its local area.
- Inquiry scan is used by devices to make themselves available to inquiring devices.
- Page state is entered by the master, which transmits paging messages to the intended slave device.
- Page scan is used by a device to allow paging devices to establish connection with it.
- Connection–Active is a state in which the slave moves to the master's frequency hop and timing sequence by switching to the master's clock.
- Connection–Hold mode allows the device to maintain its active member address while ceasing to support ACL traffic for a certain time to free bandwidth for other operations such as scanning, paging, inquiry, or low-power sleep.
- Connection–Sniff mode allows the device to listen for traffic by using a predefined slot time.
- Connection–Park mode allows the device to enter low-power sleep mode by giving up its active member address and listening for traffic only occasionally.

8.13: A connection between a host and Bluetooth device is established in the following steps:

1. Host requests an inquiry.
2. Inquiry is sent using the inquiry hopping sequence.
3. Inquiry scanning devices respond to the inquiry scan with FHS packets that contain all the information needed to connect with them.
4. The contents of the FHS packets are passed back to the host.
5. The host requests connection to one of the devices that responded to the inquiry.
6. Paging is used to initiate a connection with the selected device.
7. If the selected device is page scanning, it responds to the page.
8. If the page-scanning device accepts the connection, it will start hopping using the master's frequency hopping sequence and timing.

8.14: The host controls a Bluetooth module by using the following HCI commands:

1. Link control, for instance, setting up, tearing down, and configuring links.
2. Power-saving modes and switching of the roles between master and slave.
3. Direct access to information on the local Bluetooth module and access to information on remote devices by triggering LMP exchanges.
4. Control of baseband features, for instance, timeouts.
5. Retrieving status information on a module.
6. Invoking Bluetooth test modules for factory testing and for Bluetooth qualification.

8.15: Logical Link Control and Adaptation Protocol (L2CAP) send data packets to HCI or to LM. The functions of L2CAP include

- multiplexing between higher layer protocols that allow them to share lower layer links;
- segmentation and reassembly to allow transfer of larger packets than those that lower layers support;
- group management by providing one-way transmission to a group of other Bluetooth devices; and
- Quality-of-service management for higher layer protocols.

L2CAP provides the following facilities needed by higher layer protocols:

- Establishing links across underlying ACL channels using L2CAP signals.
- Multiplexing between different higher layer entities by assigning each one its own connection ID.
- Providing segmentation and reassembly facilities to allow large packets to be sent across Bluetooth connections.

9

Routing protocols in mobile and wireless networks

Mobile and wireless networks allow the users to access information and services electronically, regardless of their geographic location. There are infrastructured networks and infrastructureless (*ad hoc*) networks. Infrastructured network consists of a network with fixed and wired gateways. A mobile host communicates with a Base Station (BS) within its communication radius. The mobile unit can move geographically while it is communicating. When it goes out of range of one BS, it connects with a new BS and starts communicating through it by using a handoff. In this approach, the BSs are fixed.

In contrast to infrastructure-based networks, in *ad hoc* networks all nodes are mobile and can be connected dynamically in an arbitrary manner. All nodes of these networks behave as routers and take part in discovery and maintenance of routes to other nodes in the network. *Ad hoc* networks are very useful in emergency search-and-rescue operations, meetings, or conventions in which persons wish to quickly share information and data acquisition operations in inhospitable terrain.

An *ad hoc* network is a collection of mobile nodes forming a temporary network without the aid of any centralized administration or standard support services regularly available in conventional networks. We assume that the mobile hosts use wireless radio frequency transceivers as their network interface, although many of the same principles will apply to infrared and wire-based networks. Some form of routing protocol is necessary in these *ad hoc* networks since two hosts wishing to exchange packets may not be able to communicate directly.

Wireless network interfaces usually operate at significantly slower bit rates than their wire-based counterparts. Frequent flooding of packets throughout the network, a mechanism many protocols require, can consume significant portions of the available network bandwidth. *Ad hoc* routing protocols must minimize bandwidth overhead at the same time as they enable routing.

Ad hoc networks must deal with frequent changes in topology. Mobile nodes change their network location and link status on a regular basis. New nodes may unexpectedly join the network or existing nodes may leave or be turned off. *Ad hoc* routing protocols must minimize the time required to converge after the topology changes. A low convergence time is more critical in *ad hoc* networks because temporary routing loops can result in packets being transmitted in circles, further consuming valuable bandwidth.

The routing protocols meant for wired networks cannot be used for mobile *ad hoc* networks because of the mobility of networks. The *ad hoc* routing protocols can be divided into two classes: table-driven and on-demand routing, on the basis of when and how the routes are discovered. In table-driven routing protocols, consistent and up-to-date routing information to all nodes is maintained at each node, whereas in on-demand routing the routes are created only when desired by the source host. We discuss current table-driven protocols as well as on-demand protocols.

9.1 TABLE-DRIVEN ROUTING PROTOCOLS

In table-driven routing protocols, each node maintains one or more tables containing routing information to every other node in the network. All nodes update these tables so as to maintain a consistent and up-to-date view of the network. When the network topology changes, the nodes propagate update messages throughout the network in order to maintain a consistent and up-to-date routing information about the whole network. These routing protocols differ in the method by which the information regarding topology changes is distributed across the network and in the number of necessary routing-related tables.

9.1.1 Destination-sequenced distance-vector routing

Destination-Sequenced Distance-Vector Routing (DSDV) is an adoption of a conventional routing protocol to *ad hoc* networks. DSDV is based on the Routing Information Protocol (RIP) used in parts of the Internet. Consequently, DSDV only makes use of bidirectional links.

In DSDV, packets are routed between nodes of an *ad hoc* network using a Routing Table (RT) stored at each node. Each RT at each node contains a list of the addresses of every other node in the network. Along with each node's address, the table contains the address of the next hop for a packet to take in order to reach the node.

DSDV generates and maintains the RTs. Every time the network topology changes, the RT in every node needs to be updated. In addition to the destination address and next hop address, RTs maintain the route metric (the number of hops) and the route sequence number.

Periodically, or immediately when network topology changes are detected, each node will broadcast an RT update packet. The update packet starts out with a metric of one. This signifies to each receiving neighbor that it is one hop away from the node. The neighbors will increment this metric and then retransmit the update packet. This process repeats itself until every node in the network has received a copy of the update packet with a corresponding metric. If a node receives duplicate update packets, the node will only consider the update packet with the smallest metric and ignore the remaining ones.

To distinguish stale update packets from valid ones, each update packet is tagged by the original node with a sequence number. The sequence number is a monotonically increasing number that uniquely identifies each update packet from a given node. Consequently, if a node receives an update packet from another node, the sequence number must be equal to or greater than the sequence number already in the RT; otherwise the update packet is stale and ignored. If the sequence number matches the sequence number in the RT, then the metric is compared and updated.

Each time an update packet is forwarded, the packet not only contains the address of the destination but it also contains the address of the transmitting node. The address of the transmitting node is entered into the RT as the next hop. The update packets with the higher sequence numbers are always entered into the RT, regardless of whether they have a higher metric or not.

Each node must periodically transmit its entire RT to its neighbors using update packets. The neighbors will update their tables on the basis of this information, if required. Likewise, each node will listen to its neighbors update packets and update its own RT.

Mobile nodes cause broken links as they move from place to place. The broken link may be detected by the communication hardware, or it may be inferred if no broadcasts have been received for a while from a former neighbor. A broken link is described as having a metric of infinity. Since this qualifies as a substantial route change, the detecting node will broadcast an update message for the lost destination. This update message will have a new sequence number and a metric of infinity. This will essentially cause the RT entries for the lost node to be flushed from the network. Routes to the lost node will be established again when the lost node sends out its next broadcast.

To avoid nodes and their neighbors generating conflicting sequence numbers when the topology changes, nodes only generate even sequence numbers for themselves, and neighbors responding to link changes only generate odd sequence numbers.

DSDV is based on a conventional routing protocol, RIP, adopted for use in *ad hoc* networks. Routing is achieved by using RTs maintained by each node. The bulk of the complexity in DSDV is in generating and maintaining these RTs.

DSDV requires nodes to periodically transmit RT update packets, regardless of network traffic. These update packets are broadcast throughout the network so every node in the network knows how to reach every other node. As the number of nodes in the network grows, the size of the RTs and the bandwidth required to update them also grows. This overhead is DSDV's main weakness, as Broch *et al.* found in their simulations of 50-node networks. Furthermore, whenever the topology changes, DSDV is unstable until update packets propagate throughout the network. Broch found DSDV to be the most difficult in dealing with high rates of node mobility.

Every mobile station maintains an RT that lists all available destinations, the number of hops to reach the destination, and the sequence number assigned by the destination node. The sequence number is used to distinguish stale routes from new ones and thus avoid the formation of loops. The stations periodically transmit their RTs to their immediate neighbors. A station also transmits its RT if a significant change has occurred in its table from the last update sent. Thus, the update is both time-driven and event-driven. The RT updates can be sent in two ways: a full dump or an incremental update. A full dump sends the full RT to the neighbors and could span many packets, whereas in an incremental update, only those entries from the RT are sent that have a metric change since the last update, and they must fit in a packet. If there is space in the incremental update packet, then those entries may be included whose sequence number has changed. When the network is relatively stable, incremental updates are sent to avoid extra traffic and full dumps are relatively infrequent. In a fast-changing network, incremental packets can grow, and full dumps will be more frequent. Each route update packet, in addition to the RT information, also contains a unique sequence number assigned by the transmitter. The route labeled with the highest (i.e., the most recent) sequence number is used. If two routes have the same sequence number, then the route with the best metric (i.e., the shortest route) is used. On the basis of the past history, the stations estimate the settling time of routes. The stations delay the transmission of a routing update by settling time so as to eliminate those updates that would occur if a better route was found very soon.

9.1.2 The wireless routing protocol

The Wireless Routing Protocol (WRP) is a table-based distance-vector routing protocol. Each node in the network maintains a Distance table, an RT, a Link-cost table, and a Message Retransmission List (MRL). The Distance table of a node x contains the distance of each destination node y *via* each neighbor z of x. It also contains the downstream neighbor of z through which this path is realized. The RT of node x contains the distance of each destination node y from node x, the predecessor and the successor of node x on this path. It also contains a tag to identify if the entry is a simple path, a loop, or invalid. Storing predecessor and successor in the table is beneficial in detecting loops and avoiding counting-to-infinity problems. The Link-cost table contains cost of link to each neighbor of the node and the number of timeouts since an error-free message was received from that neighbor. The MRL contains information to let a node know which of its neighbors has not acknowledged its update message and to retransmit the update message to that neighbor.

Nodes exchange RTs with their neighbors using update messages periodically as well as on link changes. The nodes present on the response list of update message (formed using MRL) are required to acknowledge the receipt of the update message. If there is no change in RT since the last update, the node is required to send an idle Hello message to ensure connectivity. On receiving an update message, the node modifies its distance table and looks for better paths using new information. Any new path so found is relayed back to the original nodes so that they can update their tables. The node also updates its RT if the new path is better than the existing path. On receiving an Acknowledgement ACK, the node updates its MRL. A unique feature of this algorithm is that it checks the consistency of all its neighbors every time it detects a change in link of any of its neighbors. A consistency check in this manner helps eliminate looping situations in a better way and it also has fast convergence.

9.1.3 Global state routing

Global State Routing (GSR) is similar to DSDV. It takes the idea of link state routing but improves it by avoiding flooding of routing messages. In this algorithm, each node maintains a Neighbor list, a Topology table, a Next Hop table, and a Distance table. Neighbor list of a node contains the list of its neighbors (here all nodes that can be heard by a node are assumed to be its neighbors). For each destination node, the Topology table contains the link state information as reported by the destination and the timestamp of the information. For each destination, the Next Hop table contains the next hop to which the packets for this destination must be forwarded. The Distance table contains the shortest distance to each destination node.

The routing messages are generated on a link change as in link state protocols. On receiving a routing message, the node updates it's Topology table if the sequence number of the message is newer than the sequence number stored in the table. After this, the node reconstructs its RT and broadcasts the information to its neighbors.

9.1.4 Fisheye state routing

Fisheye State Routing (FSR) is an improvement of GSR. The large size of update messages in GSR wastes a considerable amount of network bandwidth. In FSR, each update message does not contain information about all nodes. Instead, it exchanges information about closer nodes more frequently than it does about farther nodes, thus reducing the update message size. Each node receives accurate information about neighbors, and the detail and accuracy of information decreases as the distance from the node increases. The scope is defined in terms of the nodes that can be reached in a certain number of hops. The center node has the most accurate information about the other nodes. Even though a node does not have accurate information about distant nodes, the packets are routed correctly because the route information becomes more and more accurate as the packet moves closer to the destination. FSR scales well to large networks as the overhead is controlled in this scheme.

9.1.5 Hierarchical state routing

The characteristic feature of Hierarchical State Routing (HSR) is multilevel clustering and logical partitioning of mobile nodes. The network is partitioned into clusters and a cluster head elected as in a cluster-based algorithm. In HSR, the cluster heads again organize themselves into clusters and so on. The nodes of a physical cluster broadcast their link information to each other. The cluster head summarizes its cluster's information and sends it to the neighboring cluster heads *via* the gateway. These cluster heads are members of the cluster on a level higher and they exchange their link information as well as the summarized lower-level information among themselves and so on. A node at each level floods to its lower level the information that it obtains after the algorithm has run at that level. The lower level has a hierarchical topology information. Each node has a hierarchical address. One way to assign hierarchical address is to use the cluster numbers on the way from the root to the node. A gateway can be reached from the root *via* more than one path; thus, a gateway can have more than one hierarchical address. A hierarchical address is enough to ensure delivery from anywhere in the network to the host.

In addition, nodes are also partitioned into logical subnetworks and each node is assigned a logical address <subnet, host>. Each subnetwork has a Location Management Server (LMS). All the nodes of that subnet register their logical address with the

LMS. The LMS advertise their hierarchical address to the top levels and the information is sent down to all LMS too. The transport layer sends a packet to the network layer with the logical address of the destination. The network layer finds the destination's LMS from its LMS and then sends the packet to it. The destination's LMS forwards the packet to the destination. Once the source and destination know each other's hierarchical addresses, they can bypass the LMS and communicate directly. Since logical address/hierarchical address is used for routing, it is adaptable to network changes.

9.1.6 Zone-based hierarchical link state routing protocol

In Zone-based Hierarchical Link State Routing Protocol (ZHLS), the network is divided into nonoverlapping zones. Unlike other hierarchical protocols, there is no zone-head. ZHLS defines two levels of topologies – node level and zone level. A node level topology tells how nodes of a zone are connected to each other physically. A virtual link between two zones exists if at least one node of a zone is physically connected to some node of the other zone. Zone level topology tells how zones are connected together. There are two types of Link State Packets (LSP) as well – node LSP and zone LSP. A node LSP of a node contains its neighbor node information and is propagated with the zone, whereas a zone LSP contains the zone information and is propagated globally. Each node has full node connectivity knowledge about the nodes in its zone and only zone connectivity information about other zones in the network. Given the zone id and the node id of a destination, the packet is routed on the basis of the zone id till it reaches the correct zone. Then in that zone, it is routed on the basis of the node id. A <zone id, node id> of the destination is sufficient for routing, so it is adaptable to changing topologies.

9.1.7 Cluster-head gateway switch routing protocol

Cluster-head Gateway Switch Routing (CGSR) uses as basis the DSDV Routing algorithm. The mobile nodes are aggregated into clusters and a cluster head is elected. All nodes that are in the communication range of the cluster head belong to its cluster. A gateway node is a node that is in the communication range of two or more cluster heads. In a dynamic network, a cluster-head scheme can cause performance degradation due to frequent cluster-head elections, so CGSR uses a Least Cluster Change (LCC) algorithm. In LCC, cluster-head change occurs only if a change in network causes two cluster heads to come into one cluster, or one of the nodes moves out of the range of all the cluster heads.

The general algorithm works in the following manner: the source of the packet transmits the packet to its cluster head. From this cluster head, the packet is sent to the gateway node that connects this cluster head and the next cluster head along the route to the destination. The gateway sends it to that cluster head and so on till the destination cluster head is reached in this way. The destination cluster head then transmits the packet to the destination.

Each node maintains a cluster member table that has mapping from each node to its respective cluster head. Each node broadcasts its cluster member table periodically and updates its table after receiving other node's broadcasts using the DSDV algorithm.

In addition, each node also maintains an RT that determines the next hop to reach the destination cluster.

On receiving a packet, a node finds the nearest cluster head along the route to the destination according to the cluster member table and the RT. Then it consults its RT to find the next hop in order to reach the cluster head selected in step one and transmits the packet to that node.

9.2 ON-DEMAND ROUTING PROTOCOLS

In contrast to table-driven routing protocols, all up-to-date routes are not maintained at every node; instead, the routes are created as and when they are required. When source wants to send a packet to destination, it invokes the route discovery mechanisms to find the path to the destination. The route remains valid till the destination is reachable or until the route is no longer needed.

9.2.1 Temporally ordered routing algorithm

Temporally Ordered Routing Algorithm (TORA) is a distributed routing protocol based on a link reversal algorithm. It is designed to discover routes on-demand, provide multiple routes to a destination, establish routes quickly, and minimize communication overhead by localizing the reaction to topological changes when possible. Route optimality (the shortest-path routing) is considered of secondary importance, and longer routes are often used to avoid the overhead of discovering newer routes. It is also not necessary (nor desirable) to maintain routes between every source/destination pair at all times.

The actions taken by TORA can be described in terms of water flowing downhill toward a destination node through a network of tubes that model the routing state of the network. The tubes represent links between nodes in the network, the junctions of the tubes represent the nodes, and the water in the tubes represents the packets flowing toward the destination. Each node has a height with respect to the destination that is computed by the routing protocol. If a tube between two nodes becomes blocked such that water can no longer flow through it, the height of the nodes is set to a height greater than that of any neighboring nodes, such that water will now flow back out of the blocked tube and find an alternate path to the destination.

At each node in the network, a logically separate copy of TORA is run for each destination. When a node needs a route to a particular destination, it broadcasts a route query packet containing the address of the destination. This packet propagates through the network until it reaches either the destination or an intermediate node having a route to the destination. The recipient of the query packet then broadcasts an update packet listing its height with respect to the destination (if the recipient is the destination, this height is 0). As this packet propagates back through the network, each node that receives the update sets its height to a value greater than the height of the neighbor from which the update was received. This has the effect of creating a series of directed links from the original sender of the query to the node that initially generated the update.

When a node discovers that a route to a destination is no longer valid, it adjusts its height so that it is at a local maximum with respect to its neighbors and transmits an update packet.

When a node detects a network partition, where a part of the network is physically separated from the destination, the node generates a clear packet that resets the routing state and removes invalid routes from the network.

TORA is a routing layer above network level protocol called the *Internet Mobile Ad hoc Networking* (MANET) *Encapsulation Protocol* (IMEP). IMEP is designed to support the operation of many routing algorithms, network control protocols, or other upper layer protocols intended for use in mobile *ad hoc* networks. The protocol incorporates mechanisms for supporting link status and neighbor connectivity sensing, control packet aggregation and encapsulation, one-hop neighbor broadcast reliability, multipoint relaying, network-layer address resolution, and provides hooks for interrouter authentication procedures.

In TORA, each node must maintain a structure describing the node's height as well as the status of all connected links per connection supported by the network. Each node must also be in constant coordination with neighboring nodes in order to detect topology changes and converge. As was found with DSDV, routing loops can occur while the network is reacting to a change in topology.

TORA is designed to carry IP traffic over wireless links in an *ad hoc* network. On the basis of simulation results by Park and Corson, it is best suited to large, densely packed arrays of nodes with very low node mobility. Broch *et al.* simulated node mobility and found TORA to be encumbered by its layering on top of IMEP and that IMEP caused considerable congestion when TORA was trying to converge in response to node mobility. This resulted in TORA requiring between one to two orders of magnitude more routing overhead than other *ad hoc* routing protocols investigated by Broch.

TORA is a highly adaptive, efficient, and scalable distributed routing algorithm based on the concept of link reversal. TORA is proposed for highly dynamic mobile, multihop wireless networks. It is a source-initiated on-demand routing protocol. It finds multiple routes from a source node to a destination node. The main feature of TORA is that the control messages are localized to a very small set of nodes near the occurrence of a topological change. To achieve this, the nodes maintain routing information about adjacent nodes. The protocol has three basic functions: route creation, route maintenance, and route erasure.

Each node has a quintuple associated with it:

1. Logical time of a link failure
2. The unique ID of the node that defined the new reference level
3. A reflection indicator bit
4. A propagation ordering parameter
5. The unique ID of the node.

The first three elements collectively represent the reference level. A new reference level is defined each time a node loses its last downstream link due to a link failure. The last two values define a delta with respect to the reference level.

Route creation is done using Query (QRY) and Update (UPD) packets. The route creation algorithm starts with the height (propagation ordering parameter in the quintuple) of destination set to 0 and all other node's height set to NULL (i.e., undefined). The source broadcasts a QRY packet with the destination node's identifier in it. A node with a non-NULL height responds with a UPD packet that has its height in it. A node receiving a UPD packet sets its height to one more than that of the node that generated the UPD. A node with higher height is considered upstream and a node with lower height is considered downstream. In this way, a Directed Acyclic Graph (DAG) is constructed from the source to the destination.

When a node moves, the DAG route is broken, and route maintenance is needed to reestablish a DAG for the same destination. When the last downstream link of a node fails, it generates a new reference level. This results in the propagation of that reference level by neighboring nodes. Links are reversed to reflect the change in adapting to the new reference level. This has the same effect as reversing the direction of one or more links when a node has no downstream links.

In the route erasure phase, TORA floods a broadcast clear packet (CLR) throughout the network to erase invalid routes.

In TORA, there is a potential for oscillations to occur, especially when multiple sets of coordinating nodes are concurrently detecting partitions, erasing routes, and building new routes based on each other. Because TORA uses internodal coordination, its instability problem is similar to the count-to-infinity problem in distance-vector routing protocols, except that such oscillations are temporary and route convergence will ultimately occur.

9.2.2 Dynamic source routing protocol

Dynamic Source Routing Protocol (DSRP) is designed to allow nodes to dynamically discover a source route across multiple network hops to any destination in the *ad hoc* network. When using source routing, each packet to be routed carries in its header the complete, ordered list of nodes through which the packet must pass. A key advantage of source routing is that intermediate hops do not need to maintain routing information in order to route the packets they receive, since the packets themselves already contain all the necessary routing information.

DSRP does not require the periodic transmission of router advertisements or link status packets, reducing the overhead of DSRP. In addition, DSRP has been designed to compute correct routes in the presence of unidirectional links.

Source routing is a routing technique in which the sender of a packet determines the complete sequence of nodes through which to forward the packet. The sender explicitly lists this path in the packet's header, identifying each forwarding hop by the address of the next node to which the packet is to be transmitted on its way to the destination host.

DSRP is broken down into three functional components: routing, route discovery, and route maintenance. Route discovery is the mechanism by which a node wishing to send a packet to a destination obtains a path to the destination. Route maintenance is the mechanism by which a node detects a break in its source route and obtains a corrected route.

To perform route discovery, the source node broadcasts a route request packet with a recorded source route listing. Each node that hears the route request forwards the request, if appropriate, adding its own address to the recorded source route in the packet. The route request packet propagates hop-by-hop outward from the source node until either the destination node is found or until another node is found that can supply a route to the target.

Nodes forward route requests if they are not the destination node and they are not already listed as a hop in the route. In addition, each node maintains a cache of recently received route requests and does not propagate any copies of a route request packet after the first.

All source routes learned by a node are kept (memory permitting) in a route cache, which is used to further reduce the cost of route discovery. A node may learn of routes from virtually any packet the node forwards or overhears. When a node wishes to send a packet, it examines its own route cache and performs route discovery only if no suitable source route is found.

Further, when a node receives a route request for which it has a route in its cache, it does not propagate the route request, but instead returns a route reply to the source node. The route reply contains the full concatenation of the recorded route from the source and the cached route leading to the destination.

If a route request packet reaches the destination node, the destination node returns a route reply packet to the source node with the full source to destination path listed.

Conventional routing protocols integrate route discovery with route maintenance by continuously sending periodic routing updates. If the status of a link or node changes, the periodic updates will eventually reflect the change to all other nodes, presumably resulting in the computation of new routes. However, using route discovery, there are no periodic messages of any kind from any of the mobile nodes. Instead, while a route is in use, the route maintenance procedure monitors the operation of the route and informs the sender of any routing errors.

If a node along the path of a packet detects an error, the node returns a route error packet to the sender. The route error packet contains the addresses of the nodes at both ends of the hop in error. When a route error packet is received or overheard, the hop in error is removed from any route caches and all routes that contain this hop must be truncated at that point.

There are many methods of returning a route error packet to the sender. The easiest of these, which is only applicable in networks that only use bidirectional links, is to simply reverse the route contained in the packet from the original host. If unidirectional links are used in the network, the DSRP presents several alternative methods of returning route error packets to the sender.

Route maintenance can also be performed using end-to-end acknowledgments rather than the hop-by-hop acknowledgments described above. As long as a route exists by which the two end hosts can communicate, route maintenance is possible. In this case, existing transport or application level replies or acknowledgments from the original destination, or explicitly requested network level acknowledgments, may be used to indicate the status of the node's route to the other node.

Two sources of bandwidth overhead in DSRP are route discovery and route maintenance. They occur when new routes need to be discovered or when the network topology changes. However, this overhead can be reduced by employing intelligent caching techniques in each node at the expense of memory and CPU resources. The remaining source of bandwidth overhead is the required source route header included in every packet.

Broch *et al.* found DSRP to perform the best of four protocols simulated. However, Broch simulated networks of 50 nodes. As the node count increases, the detrimental effects of route discovery and maintenance can be expected to grow.

DSRP is a source-routed on-demand routing protocol. A node maintains route caches containing the source routes that it is aware of. The node updates entries in the route cache as and when it learns about new routes.

The two major phases of the protocol are route discovery and route maintenance. When the source node wants to send a packet to a destination, it looks up its route cache to determine if it already contains a route to the destination. If it finds that an unexpired route to the destination exists, then it uses this route to send the packet. But if the node does not have such a route, then it initiates the route discovery process by broadcasting a route request packet. The route request packet contains the address of the source, the destination, and a unique identification number. Each intermediate node checks whether it knows of a route to the destination. If it does not, it appends its address to the route record of the packet and forwards the packet to its neighbors. To limit the number of route requests propagated, a node processes the route request packet only if it has not already seen the packet and its address is not present in the route record of the packet.

A route reply is generated when either the destination or an intermediate node with current information about the destination receives the route request packet. A route request packet reaching such a node already contains, in its route record, the sequence of hops taken from the source to this node.

As the route request packet propagates through the network, the route record is formed. If the route reply is generated by the destination, then it places the route record from route request packet into the route reply packet. On the other hand, if the node generating the route reply is an intermediate node, then it appends its cached route to destination to the route record of route request packet and puts that into the route reply packet. To send the route reply packet, the responding node must have a route to the source. If it has a route to the source in its route cache, it can use that route. The reverse of route record can be used if symmetric links are supported. In case symmetric links are not supported, the node can initiate route discovery to source and piggyback the route reply on this new route request.

DSRP uses two types of packets for route maintenance: Route Error packet and Acknowledgements. When a node encounters a fatal transmission problem at its data link layer, it generates a Route Error packet. When a node receives a Route Error packet, it removes the hop in error from its route cache. All routes that contain the hop in error are truncated at that point. Acknowledgment packets are used to verify the correct operation of the route links. This also includes passive acknowledgments in which a node hears the next hop forwarding the packet along the route.

9.2.3 Cluster-based routing protocol

In Cluster-Based Routing Protocol (CBRP), the nodes are divided into clusters. To form the cluster, the following algorithm is used. When a node comes up, it enters the undecided state, starts a timer, and broadcasts a 'hello' message. When a cluster head gets this hello message, it responds with a triggered hello message immediately. When the undecided node gets this message, it sets its state to member. If the undecided node times out, then it makes itself the cluster head if it has a bidirectional link to some neighbor, otherwise it remains in undecided state and repeats the procedure again. Cluster heads are changed as infrequently as possible.

Each node maintains a neighbor table. For each neighbor, the neighbor table of a node contains the status of the link (uni- or bidirectional) and the state of the neighbor (cluster head or member). A cluster head keeps information about the members of its cluster and also maintains a cluster adjacency table that contains information about the neighboring clusters. For each neighbor cluster, the table has entry that contains the cluster head and the gateway through which the cluster can be reached.

When a source has to send data to destination, it floods route request packets (but only to the neighboring cluster heads). On receiving the request, a cluster head checks to see if the destination is in its cluster. If it is, then the cluster head sends the request directly to the destination, else it sends the request to all its adjacent cluster heads. The cluster-heads address is recorded in the packet so a cluster head discards a request packet that it has already seen. When the destination receives the request packet, it replies back with the route that had been recorded in the request packet. If the source does not receive a reply within a time period, it backs off exponentially before trying to send the route request again.

In CBRP, routing is done using source routing. It also uses route shortening, that is, on receiving a source route packet, the node tries to find the farthest node in the route that is its neighbor (this could have happened because of a topology change) and sends the packet to that node, thus reducing the route. If a node detects a broken link, while forwarding the packet, it sends back an error message to the source and then uses local repair mechanism. In local repair mechanism, when a node finds that the next hop is unreachable, it checks to see if the next hop can be reached through any of its neighbors or if the hop after the next hop can be reached through any other neighbor. If any one of the two works, the packet can be sent out over the repaired path.

The CBRP is a variation of the DSRP protocol. CBRP groups nodes located physically close together into clusters. Each cluster elects a cluster head. The cluster head maintains complete knowledge of cluster membership and intercluster members (called *gateways*), which have access to neighboring clusters.

The main difference between DSRP and CBRP is during the route discovery operation. DSRP floods route query packets throughout the entire network. CBRP takes advantage of its cluster structure to limit the scope of route query flooding. In CBRP, only the cluster heads (and corresponding gateways) are flooded with query packets, reducing the bandwidth required by the route discovery mechanism.

CBRP depends on nodes transmitting periodic hello packets; a large part of the gains made by DSRP are because DSRP does not require periodic packets of any kind.

9.2.4 *Ad hoc* on-demand distance-vector routing

Ad hoc On-demand Distance-Vector Routing (AODV) is an improvement on the DSDV algorithm. AODV minimizes the number of broadcasts by creating routes on-demand as opposed to DSDV that maintains the list of all the routes. To find a path to the destination, the source broadcasts a route request packet. The neighbors in turn broadcast the packet to their neighbors till it reaches an intermediate node that has a recent route information about the destination or till it reaches the destination. A node discards a route request packet that it has already seen. The route request packet uses sequence numbers to ensure that the routes are loop-free and to make sure that if the intermediate nodes reply to route requests, they reply with the latest information only.

When a node forwards a route request packet to its neighbors, it also records in its tables the node from which the first copy of the request came. This information is used to construct the reverse path for the route reply packet. AODV uses only symmetric links because the route reply packet follows the reverse path of route request packet. As the route reply packet traverses back to the source, the nodes along the path enter the forward route into their tables.

If the source moves, then it can reinitiate route discovery to the destination. If one of the intermediate nodes move, then the moved nodes neighbor realizes the link failure and sends a link failure notification to its upstream neighbors and so on till it reaches the source upon which the source can reinitiate route discovery if needed.

AODV routing is essentially a combination of both DSRP and DSDV. It borrows the basic on-demand mechanism of route discovery and route maintenance from DSRP, plus the use of hop-by-hop routing, sequence numbers, and periodic update packets from DSDV.

The main benefit of AODV over DSRP is that the source route does not need to be included with each packet. This results in a reduction of routing protocol overhead. AODV requires periodic updates that, based on simulations by Broch, consume more bandwidth than is saved from not including source route information in the packets.

9.2.5 Signal stability-based adaptive routing

Signal Stability-Based Adaptive Routing (SSA) is a variant of the AODV protocol to take advantage of information available at the link level. Both the signal quality of links and link congestion are taken into consideration when finding routes. It is assumed that links with strong signals will change state less frequently. By favoring these strong signal links in route discovery, it is hoped that routes will survive longer and the number of route discovery operations will be reduced. Link signal strength is measured when the nodes transmit periodic hello packets.

One important difference of SSA from AODV or DSRP is that paths with strong signal links are favored over optimal paths. While this may make routes longer, it is hoped that discovered routes will survive longer.

Signal Stability-based adaptive Routing protocol (SSR) is an on-demand routing protocol that selects routes on the basis of the signal strength between nodes and a node's location stability. This route selection criterion has the effect of choosing routes that have stronger connectivity. SSR is composed of two cooperative protocols: the Dynamic Routing Protocol (DRP) and the Static Routing Protocol (SRP). The DRP maintains the Signal Stability Table (SST) and RT. The SST stores the signal strength of neighboring nodes obtained by periodic beacons from the link layer of each neighboring node. Signal strength is either recorded as a strong or weak channel. All transmissions are received by DRP and processed. After updating the appropriate table entries, the DRP passes the packet to the SRP.

The SRP passes the packet up the stack if it is the intended receiver. If not, it looks up the destination in the RT and forwards the packet. If there is no entry for the destination in the RT, it initiates a route search process to find a route. Route request packets are forwarded to the next hop only if they are received over strong channels and have not been previously processed (to avoid looping). The destination chooses the first arriving route search packet to send back, as it is very likely that the packet arrived over the shortest and/or least congested path. The DRP reverses the selected route and sends a route reply message back to the initiator of route request. The DRP of the nodes along the path update their RTs accordingly.

Route search packets arriving at the destination have necessarily arrived on the path of strongest signal stability because the packets arriving over a weak channel are dropped at intermediate nodes. If the source times out before receiving a reply, then it changes the preference PREF field in the header to indicate that weak channels are acceptable, since these may be the only links over which the packet can be propagated.

When a link failure is detected within the network, the intermediate nodes send an error message to the source indicating which channel has failed. The source then sends an erase message to notify all nodes of the broken link and initiates a new route search process to find a new path to the destination.

9.2.6 Associativity-based routing

The Associativity-Based Routing (ABR) protocol defines a routing metric known as the degree of association stability. It is free from loops, deadlock, and packet duplicates. In ABR, a route is selected on the basis of the associativity states of nodes. The routes thus selected are likely to be long-lived. All nodes generate periodic beacons to signify their existence. When a neighbor node receives a beacon, it updates its associativity tables. For every beacon received, a node increments its associativity tick with respect to the node from which it received the beacon.

Association stability means connection stability of one node with respect to another node over time and space. A high value of associativity tick with respect to a node indicates a low state of node mobility, while a low value of associativity tick may indicate a high state of node mobility. Associativity ticks are reset when the neighbors of a node or the node itself move out of proximity. The fundamental objective of ABR is to find longer-lived routes for *ad hoc* mobile networks. The three phases of ABR are route discovery, Route Reconstruction (RRC), and route deletion.

The route discovery phase is a Broadcast Query (BQ) and await-reply (BQ-REPLY) cycle. The source node broadcasts a BQ message in search of nodes that have a route to the destination. A node does not forward a BQ request more than once. On receiving a BQ message, an intermediate node appends its address and its associativity ticks to the query packet. The next succeeding node erases its upstream node neighbors' associativity tick entries and retains only the entry concerned with itself and its upstream node. Each packet arriving at the destination will contain the associativity ticks of the nodes along the route from source to the destination. The destination can now select the best route by examining the associativity ticks along each of the paths. If multiple paths have the same overall degree of association stability, the route with the minimum number of hops is selected. Once a path has been chosen, the destination sends a REPLY packet back to the source along this path. The nodes on the path that the REPLY packet follows mark their routes as valid. All other routes remain inactive, thus avoiding the chance of duplicate packets arriving at the destination.

RRC phase consists of partial route discovery, invalid route erasure, valid route updates, and new route discovery, depending on which node(s) along the route move. Source node movement results in a new BQ-REPLY process because the routing protocol is source-initiated. The Route Notification (RN) message is used to erase the route entries associated with downstream nodes. When the destination moves, the destination's immediate upstream node erases its route. A Localized Query (LQ [H]) process, in which H refers to the hop count from the upstream node to the destination, is initiated to determine if the node is still reachable. If the destination receives the LQ packet, it selects the best partial route and REPLYs; otherwise, the initiating node times out and backtracks to the next upstream node. An RN message is sent to the next upstream node to erase the invalid route and to inform this node that it should invoke the LQ [H] process. If this process results in backtracking more than halfway to the source, the LQ process is discontinued and the source initiates a new BQ process.

When a discovered route is no longer needed, the source node initiates a Route Delete (RD) broadcast. All nodes along the route delete the route entry from their RTs. The RD message is propagated by a full broadcast as opposed to a directed broadcast because the source node may not be aware of any route node changes that occurred during RRCs.

9.2.7 Optimized link state routing

Optimized Link State Routing (OLSR) is a link state routing protocol. OLSR is an adoption of conventional routing protocols to work in an *ad hoc* network on top of IMEP.

The novel attribute of OLSR is its ability to track and use multipoint relays. The idea of multipoint relays is to minimize the flooding of broadcast messages in the network by reducing/optimizing duplicate retransmissions in the same region. Each node in the network selects a set of nodes in its neighborhood that will retransmit its broadcast packets. This set of selected neighbor nodes is called *the multipoint relays of that node*. Each node selects its multipoint relay set in a manner to cover all the nodes that are two hops away from it. The neighbors that are not in the multipoint relay set still receive and process broadcast packets, but do not retransmit them.

9.2.8 Zone routing protocol

The Zone Routing Protocol (ZRP) is a hybrid of DSRP, DSDV, and OLSR. In ZRP, each node proactively maintains a zone around itself using a protocol such as DSDV. The zone consists of all nodes within a certain number of hops, called the *zone radius*, away from the node. Each node knows the best way to reach each of the other nodes within its zone. The nodes that are on the edges of the zone (i.e., are exactly zone radius hops from the node) are called *border nodes* and are employed in a similar fashion to multipoint relays in OLSR. When a node needs to route a packet, it first checks to see if the destination node is within its zone. If it is, the node knows exactly how to route to the destination. Otherwise, a route search similar to DSRP is employed. However, to reduce redundant route search broadcasts, nodes only transmit route query packets to the border nodes. When the border nodes receive the search query packet, they repeat the process for their own zones.

Because ZRP only employs proactive network management in a local zone, the overhead is reduced over protocols like DSDV. When route discovery procedures are employed as in DSRP, the overhead is reduced by limiting the query packet broadcasts to border nodes.

9.2.9 Virtual subnets protocol

The Virtual Subnet Protocol (VSP) breaks up a large body of nodes into smaller logical groups called *subnets*. It then applies a hierarchical addressing scheme to these subnets. A novel routing scheme is then employed to enable broadcasting within subnets and limited broadcasting between subnets. The virtual subnet-addressing scheme is somewhat reminiscent of that used in ATM.

In this method, network nodes are assigned addresses depending on their current physical connectivity. We assume that the network is segmented into physical subnets containing mobile nodes. Each node in the network is assigned a unique address constructed of two parts: one part is a subnet address allocated to the entire subnet (subnet_id) and the other part is an address that is unique within the node's subnet (node_id).

Each node in this topology is affiliated with nodes whose address differs only in one digit; that is, node x1.x0 is affiliated with nodes x1.x0 and x1.x0. Thus, every node is affiliated with every node within its subnet, as well as one node in every other subnet. These cross-linked affiliations are the building blocks of the *ad hoc* network.

Each node in the network is affiliated with a physical subnet (the local nodes all sharing the same subnet_id) and a virtual subnet (the nodes all sharing the same node_id). Nodes that are members of a physical subnet (subnet_id) are within close proximity in a local geographic area. Nodes that are members of a virtual subnet (node_id) form a regional network (i.e., beyond a local area). All nodes within a physical subnet have the same subnet_id, while all nodes within a virtual subnet have the same node_id.

A node becomes a member of a physical subnet by acquiring the first available address (with the lowest node_id) in that subnet. Once a node becomes affiliated with a specific physical subnet, it automatically becomes a member of a virtual subnet defined by the node_id in its address. As long as a node remains within hearing distance of its subnet neighbors, it will keep its current physical subnet affiliation and its address.

When a node moves to a new location in which it cannot establish a connection with its previous physical subnet's members, it will drop its previous address and join a new physical subnet.

In the simple case in which the destination node is within two hops of the source node, packets traverse one network address digit at a time in fixed order. For example, when the source node address is 13.33 and the destination node address is 11.36, the packet would follow the route: 13.33 to 13.36 to 11.36. In this case, routing requires at most two hops.

In general, the network will be arranged such that more than two hops are necessary from source to destination. In this case, the routing is performed in two phases. In the first phase, routing is performed only in the physical subnet. Packets are routed to the node belonging to the same virtual subnet as the destination. Using the same example as above, Phase 1 consists of routing packets from 13.33 to 13.36.

In Phase 2, packets are routed between virtual subnets. Adjustments of transmission frequencies, transmission power, and/or directional antennae to facilitate logical network connections are needed. It is assumed that all nodes are capable of reaching neighboring physical subnets when required to do so.

The VSP is a method to optimize throughput when multiple frequencies and/or spatial reuse is possible, on the condition that nodes are close together relative to their transmitter range.

9.3 SUMMARY

Routing protocols for *ad hoc* networks can be divided into two categories: table-driven and on-demand routing, on the basis of when and how the routes are discovered. In table-driven routing protocols, consistent and up-to-date routing information to all nodes is maintained at each node, whereas in on-demand routing, the routes are created only when desired by the source host.

In table-driven routing protocols, each node maintains one or more tables containing routing information with every other node in the network. All nodes update these tables so as to maintain a consistent and up-to-date view of the network. When the network topology changes, the nodes propagate update messages throughout the network in order to maintain a consistent and up-to-date routing information about the whole network. These routing protocols differ in the method by which the information regarding topology changes is distributed across the network and in the number of necessary routing-related tables.

In on-demand routing protocols, all up-to-date routes are not maintained at every node; instead the routes are created as and when they are required. When source wants to send a packet to destination, it invokes the route discovery mechanisms to find the path to the destination. The route remains valid till the destination is reachable or until the route is no longer needed.

PROBLEMS TO CHAPTER 9

Routing protocols in mobile and wireless networks

Learning objectives

After completing this chapter, you are able to

- demonstrate an understanding of routing protocols in mobile and wireless networks;
- explain table-driven routing protocols; and
- explain on-demand routing protocols.

Practice problems

9.1: What is an *ad hoc* network?
9.2: What are the two categories of routing protocols for *ad hoc* networks?
9.3: What are the functions of table-driven routing protocols?
9.4: What are the functions of on-demand routing protocols?

Practice problem solutions

9.1: In an *ad hoc* network, all nodes are mobile and can be connected dynamically in an arbitrary manner. All nodes of the network behave as routers and take part in discovery and maintenance of routes to other nodes in the network. Mobile nodes change their network location and link status on a regular basis. New nodes may unexpectedly join the network or existing nodes may leave or be turned off. *Ad hoc* routing protocols must minimize the time required to converge after the topology changes. A low convergence time is more critical in *ad hoc* networks because temporary routing loops can result in packets being transmitted in circles, further consuming valuable bandwidth.

9.2: Routing protocols for *ad hoc* networks can be divided into two categories: table-driven and on-demand routing on the basis of when and how the routes are discovered. In table-driven routing protocols, consistent and up-to-date routing information to all nodes is maintained at each node, whereas in on-demand routing, the routes are created only when desired by the source host.

9.3: In table-driven routing protocols, each node maintains one or more tables containing routing information with every other node in the network. All nodes update these tables so as to maintain a consistent and up-to-date view of the network. When the network topology changes, the nodes propagate update messages throughout the network in order to maintain a consistent and up-to-date routing information about the entire network. These routing protocols differ in the method by which the topology changes information is distributed across the network and in the number of necessary routing-related tables.

9.4: In on-demand routing protocols, all up-to-date routes are not maintained at every node; instead the routes are created as and when they are required. When source wants to send a packet to destination, it invokes the route discovery mechanisms to find the path to the destination. The route remains valid till the destination is reachable or until the route is no longer needed.

10

Handoff in mobile and wireless networks

Wireless data services use small-coverage high-bandwidth data networks such as IEEE 802.11 whenever they are available and switch to an overlay service such as the General Packet Radio Service (GPRS) network with low bandwidth when the coverage of a Wireless Local Area Network (WLAN) is not available.

From the service point of view, Asynchronous Transfer Mode (ATM) combines both the data and multimedia information into the wired networks while scaling well from backbones to the customer premises networks. In Wireless ATM (WATM) networks, end user devices are connected to switches *via* wired or wireless channels. The switch is responsible for establishing connections with the fixed infrastructure network component, either through a wired or a wireless channel. A mobile end user establishes a Virtual Circuit (VC) to communicate with another end user (either mobile or ATM end user). When the mobile end user moves from one Access Point (AP) to another AP, a handoff is required. To minimize the interruption of cell transport, an efficient switching of the active VCs from the old data path to the new data path is needed. Also, the switching should be fast enough to make the new VCs available to the mobile users.

When the handoff occurs, the current QoS may not be supported by the new data path. In this case, a negotiation is required to set up new QoS. Since a mobile user may be in the access range of several APs, it will select the AP that provides the best QoS.

During the handoff, an old path is released and then a new path is established. For the mobility feature of a mobile ATM, routing of signaling is slightly different from that of the wired ATM network. First, mapping of Mobile Terminal (MT) routing identifiers to paths in the network is necessary. Also, rerouting is needed to reestablish connection when the mobiles move around. It is one of the most important challenges to reroute ongoing connections to/from mobile users as those users move among Base Stations (BSs). Connection rerouting schemes must exhibit low handoff latency, maintain efficient routes, and limit disruption to continuous media traffic while minimizing reroute updates to the network switches.

Limiting handoff latency is essential, particularly in microcellular networks where handoffs may occur frequently and users may suddenly lose contact with the previous wireless AP. To reduce the signaling traffic and to maintain an efficient route may lead to disruptions in service to the user that is intolerable for continuous media applications such as packetized audio and video. Thus, it is important to achieve a suitable trade-off between the goals of reducing signaling traffic, maintaining an efficient route, and limiting disruption to continuous media traffic, while at the same time maintaining low handoff latency. Connection rerouting procedures for ATM-based wireless networks have been proposed for performing connection rerouting during handoff.

Break-Make and Make-Break schemes are categorized as optimistic schemes because their goals are to perform simple and fast handoff with the optimistic view that disruption to user traffic will be minimal. The Crossover Switch (COS) simply reroutes data traffic through a different path to the new BS, with the connection from the source to the COS remaining unmodified. In the make-break scheme, a new translation table entry in the ATM switch (make) is created and later the old translation entry (break) is removed. This results in cells being multicast from the COS to both the new and the old BSs for a short period of time during the handoff process.

The key idea of Predictive Approaches is to predict the next BS of the mobile endpoint and perform advance multicasting of data to the BS. This approach requires the maintenance of multiple connection paths to many or all the neighbors of the current BS of the mobile endpoint.

The basic idea of chaining approaches to connection rerouting is to extend the connection from the old to the new BS in the form of a chain. Chaining results in increased end-to-end delay and less efficient routing of the connection.

Chaining, followed by the make-break scheme, which involved a real-time handoff using the chaining scheme and, if necessary, a non-real-time rerouting using the make-break scheme, shows good performance in connection rerouting, because the separation of the real-time nature of handoffs and efficient route identification in this scheme allows it to perform handoffs quickly, and, at the same time, maintains efficient routes in the fixed part of the network.

The main development in shaping up the future high-speed (gigabit) networking is the emergence of Broadband ISDN (B-ISDN) and ATM. With its cell switching and the support of Virtual Path (VP) and Virtual Circuit (VC), ATM can provide a wide variety of traffic and diverse services, including real-time multimedia (data, voice, and video) applications. Because of its efficiency and flexibility, ATM is considered the most promising transfer technique for the implementation of B-ISDN, and for the future of high-speed wide and local area networks.

Handoff is important in any mobile network because of the default cellular architecture employed to maximize spectrum utilization. When a Mobile Terminal moves away from a BS, the signal level degrades, and there is a need to switch communications to another BS. Handoff is the mechanism by which an ongoing connection between an MT or host (MH) and a correspondent terminal or host (CH) is transferred from one point of access to the fixed network, and to another. In cellular voice telephony and mobile data networks, such points of attachment are referred to as base stations and in WLANs they are called *access points*. In either case, such a point of attachment serves a coverage area called a *cell.* Handoff, in the case of cellular telephony, involves the transfer of voice call from one BS to another. In the case of WLANs, it involves transferring the connection from one AP to another. In hybrid networks, it will involve the transfer of a connection from one BS to another, from an AP to another, between a BS and an AP, or *vice versa.*

WATM networks are typically inter-networked with a wired network (an ATM network) that provides wired connectivity among BSs in the wireless network, as well as connectivity to other fixed endpoints. In Figure 10.1, the service area in a wireless network is partitioned into cells. A cell is the region that receives its wireless coverage from a single BS. In a typical scenario, the coverage of the cells overlaps and the BSs are connected to each other and to fixed endpoints (e.g., hosts) through a wired ATM-based backbone network. A route connects a mobile device to a fixed endpoint.

In Figure 10.1, the Control and Switching Unit (CSU) provides mobility-related signaling (registration, deregistration, location update, and handoff), as well as routing of ATM cells. It is assumed that the CSU incorporates a typical commercially available ATM switch. The operation of the CSU is supported by a specially designed database (DB).

For a voice user, handoff results in an audible click interrupting the conversation for each handoff, and because of handoff, data users may lose packets and unnecessary congestion control measures may degrade the signal level; however, it is a random process, and simple decision mechanisms such as those based on signal strength measurements result in the ping-pong effect. The ping-pong effect refers to several handoffs that occur back and forth between two BSs. This takes a severe toll on both the user's quality perception and the network load. One way of eliminating the ping-pong effect is to persist with a BS for as long as possible. However, if handoff is delayed, weak signal reception persists unnecessarily, resulting in lower voice quality, increasing the probability of call drops and/or degradation of quality of service (QoS). Consequently, more complex algorithms are needed to decide on the optimal time for handoff.

While significant work has been done on handoff mechanisms in circuit-switched mobile networks, there is not much literature available on packet-switched mobile networks.

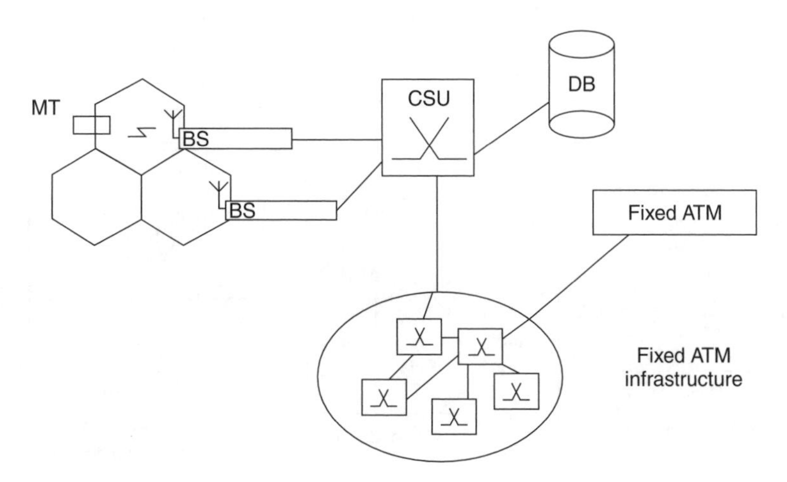

Figure 10.1 Configuration of WATM network.

Performance measures such as call blocking and call dropping are applicable only to real-time traffic and may not be suitable for the bursty traffic that exists in client-server applications. When a voice call is in progress, allowed latency is very limited, resource allocation has to be guaranteed, and, while occasionally some packets may be dropped and moderate error rates are permissible, retransmissions are not possible, and connectivity has to be maintained continuously. On the other hand, bursty data traffic by definition needs only intermittent connectivity, and it can tolerate greater latencies and employ retransmission of lost packets. In such networks, handoff is warranted only when the terminal moves out of coverage of the current point of attachment, or the traffic load is so high that a handoff may result in greater throughput and utilization.

10.1 SIGNALING HANDOFF PROTOCOL IN WATM NETWORKS

Signaling is a problem area in WATM networks. Apart from the conventional signaling solutions encountered in wired networks, additional signaling is needed to cover the mobility requirements of terminals. Wired ATM networks, which are enjoying *commercial* growth, do not support mobility of user terminal equipment. A possible solution to this problem is the integration of the required mobility extensions with the standard signaling protocols.

Protocol stacks in WATM are shown in Figure 10.2. This protocol includes mobility function for handoff. In Figure 10.2, we have the following components:

- *MMC*: Mobility Management and Control
- *RRM*: Radio Resource Manager
- *SAAL*: Signaling ATM Adaptation Layer
- *CCS*: Call Control and Signaling
- *UNI*: User-Network Interface

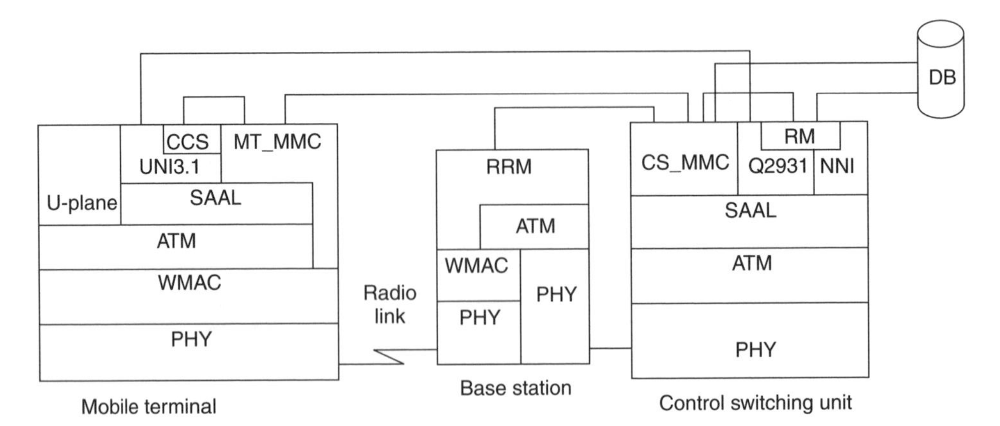

Figure 10.2 Protocol stacks.

- *WMAC*: Wireless Medium Access Control
- *S-channels*: Permanent Virtual Circuits (PVCs) intended for standard signal
- *M-channels*: PVCs intended for mobility signaling
- *U-plane*: User plane.

In Figure 10.2, the standard signaling is left unaffected. To support mobility functions, the only modifications added to the existing infrastructure are the new interfaces with the controlling entities of standard signaling (i.e., CCS, resource manager). In terms of module-entity instances, there is a one-to-one mapping between the RRM and the BSs (each BS has an RRM instance). There is also a one-to-one relationship between active MTs and MMC instances residing within the MMC in the CCS entity. In each MT, only one MMC and one CCS instance are needed.

The CS_MMC module is responsible for handling all mobility-related procedures (i.e., handover, registration, and location update) on the network side. Specifically, the CS_MMC deals with the following tasks:

- the establishment of the M-channel through which the mobility-related messages are exchanged;
- the coordination of wireless and fixed resources, during the execution of mobility and standard signaling procedures;
- the switching of signaling and data connections whenever an MT crosses the boundaries of a cell;
- the updating of the location of an MT in the CSU-hosted DB.

The basic steps involved in handoff occur in an application scenario, involving Mobile Multi-User Platforms (MMUPs) equipped with (onboard) private ATM networks.

Connection handoff is the procedure of rerouting an existing connection from the previous AP to the next when a mobile moves across a cell boundary. Success rate of handoffs and their smooth completion are crucial to providing satisfactory quality of service to mobile users. A handoff is successful if the connection is reestablished with the MMUP in the new cell. A handoff is smooth if the connection suffers no or minimum perceivable disruption during the transfer. Smoothness of handoffs depends on the number of connections requiring handoff, and the time between initiation of a handoff and loss of contact with the previous AP. MMUPs can have a large number of connections existing simultaneously owing to the presence of multiple users onboard, and a short time period available for handoffs because of high travel speeds.

10.2 CROSSOVER SWITCH DISCOVERY

The basic step common to most handoff schemes for mobile ATM networks is crossover switch discovery for each connection that required a handoff. A crossover switch (COS) is an intermediate switch along the current path of a connection that has nonoverlapping paths to both the current and the next APs. The process of selecting a COS for a connection can be initiated at the previous or next AP. Once a particular COS is selected, appropriate resources for the connection are procured along the new subpath (between the COS

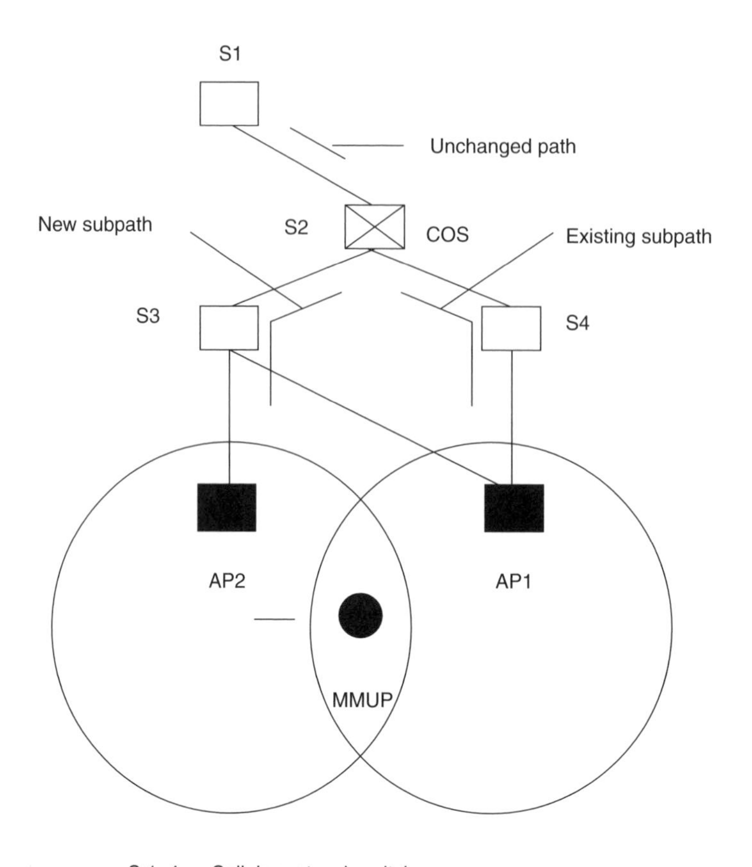

Figure 10.3 Crossover switch–based connection rerouting during handoff.

and the new AP). After the COS starts forwarding packets onto the new subpath, the existing subpath to the previous AP is torn down, thereby completing the handoff of that connection. Figure 10.3 shows an example of COS rerouting of a connection during handoff. S2 is the COS to be found during the handoff process. In Figure 10.3, we have

- S1, S2, S3, and S4, which are the cellular network switches
- COS, which is the crossover switch
- AP1, and AP2, which are the access points.

The selection of a particular switch as a COS for a connection depends on several factors, including

- switch capability
- selection policy

- new access point
- previous path.

The only selection parameter that differs among connections to a particular MMUP is the previous path, which depends on the other endpoint of each connection.

10.3 REROUTING METHODS

In ATM-based wireless networks, fast, seamless, and distributed handoff is a critical issue. Handoff call management plays a major role in supporting acceptable levels of QoS in Personal Communication Systems.

Some of the important concerns in performing such connection rerouting are

- limiting handoff latency;
- maintaining an efficient route;
- limiting disruption of continuous media traffic;
- limiting network switch update rates due to rerouting.

Limiting handoff latency is essential, particularly in microcellular networks where handoffs may occur frequently, and users may suddenly lose contact with the previous wireless AP. The process of maintaining an efficient route can also potentially lead to disruptions in user traffic that are intolerable for continuous media applications such as packetized audio and video. Thus, it is important to achieve a suitable trade-off between the goals of maintaining an efficient route and limiting disruption to continuous media traffic while maintaining low handoff latency at the same time. In order to not overload the switch, this must be done while keeping the switch updates due to connection rerouting, low.

Connection rerouting procedures for ATM-based wireless networks include handoff schemes, which are Switched Virtual Circuit (SVC)–based and PVC-based schemes.

Connection rerouting involves the location of the COS. A COS is defined to be the farthest switch from the fixed endpoint that is also the point of divergence between the new and old routes connecting the mobile and fixed endpoint.

Four general approaches toward connection rerouting are proposed:

- Optimistic handoff approach
- Ordered handoff approach
- Predictive handoff approach
- Chaining handoff approach.

The goal of an optimistic handoff scheme is to perform simple and fast handoffs with the optimistic view that disruption to user traffic will be minimal. The COS simply reroutes data traffic through a different path to the new BS with the connection from the source to the COS remaining unmodified.

The goal of an ordered approach is to provide ordered lossless data delivery during handoffs. The incremental and multicast-based rerouting schemes fall into this category.

However, complex protocols with resynchronization mechanisms and buffering at the BS are necessary to ensure lossless connection rerouting.

In predictive approaches to connection rerouting, the key idea is to predict the next BS of the mobile endpoint and perform advance multicasting of data to the BS. This approach requires the maintenance of multiple connection paths to many or all the neighbors of the current BS of the mobile endpoint.

10.4 OPTIMIZED COS DISCOVERY THROUGH CONNECTION GROUPING

Among a large number of MMUP connections, groups of connections going to the same external host can occur naturally. Within each such group, all connections will probably share the same path between the MMUP and their common external host owing to the limited number of border switches within the MMUP network. A single common COS can help reroute all connections within such a group at the time of a handoff. Depending on the size of such groups, a single COS discovery per group (instead of per connection) can cut down the total time required for handoffs significantly. Since connection handoffs are performed within the confines of the cellular network, connections to different external hosts that share a common subpath within the cellular network can be grouped together as well.

Since hop-by-hop path information, which is necessary to perform such grouping of connections, is not accumulated during connection setup, switches within the cellular network are updated to run a modified Private Network to Network Interface (PNNI) protocol that accumulates and forwards hop-by-hop path information to the MMUP during connection setup. For connections originating at a host external to the MMUP, a path list is created at the first gateway switch (in the cellular network) encountered during connection setup and is eventually passed onto a border switch within the MMUP. Each border switch in the MMUP network maintains a group database wherein connections sharing a common subpath are placed together. Each intermediate switch that receives a path list simply appends its identifier and forwards it to the next hop on the path. For connections originating within the MMUP, the path list is created at the access point on the path and forwarded up to a gateway switch, which returns the list to the MMUP along the partial path established using a special signaling message. Intermediate nodes are required to simply forward any such incoming messages onto the next hop along the path. When presented with subpath information at the time of a connection setup, a border switch within the MMUP groups connections according to the commonality of their subpaths within the cellular network. At the time of a handoff, the border switch passes on the group information to the access point, which is responsible for initiating COS discovery. A list of VCs that belong to a common group accompanies the group-COS discovery request so that path state can be setup and resources procured individually for each individual connection.

10.5 SCHEDULE-ASSISTED HANDOFFS

Preplanned travel schedules can be used to improve smoothness of handoffs in high-speed MMUP application scenarios. A schedule provides the MMUP with information about the upcoming cell in advance of its intercell moves. Consequently, an MMUP can trigger COS discoveries for existing connections a short time before it establishes contact with the next AP. This time period should be determined individually for each application scenario on the basis of the specific system characteristics and trial observations and should be set to the necessary minimum. Advance COS discoveries are not needed if cell overlap regions are large enough.

When the MMUP is close enough to the next AP for the mechanism to be triggered, it initiates a COS discovery for some or all existing connections (or groups) through the current AP. The COS-discovery process results in establishment of new subpaths from the COS to the next AP. These connections are maintained if a call proceeding sends signaling to the next AP until the MMUP establishes contact with it (or the timer expires), upon which all pending connection requests are forwarded onto the MMUP. After the MMUP confirms successful reestablishment of the connections, the COS begins switching data along the new paths and initiates tear down of the old subpaths.

Since connection resources are held along the new subpaths until the MMUP makes the move to the next cell, it is important to keep the advance trigger threshold to the necessary minimum. Nevertheless, some connections that are handed off early may terminate before the actual move is made. They are rejected by the MMUP at the next AP, and the corresponding new subpaths are torn down.

10.6 HANDOFF IN LOW EARTH ORBIT (LEO) SATELLITE NETWORKS

A handover rerouting algorithm, referred to as Footprint Handover Rerouting Protocol (FHRP), has been proposed to handle the intersatellite handover problem. The protocol addresses the trade-off between the simplicity of the partial connection rerouting and the optimality of the complete rerouting. The FHRP is a hybrid algorithm that consists of the augmentation and the Footprint Rerouting (FR) phases. In the augmentation phase, a direct link from the new end satellite to the existing connection routes is found. This way the route can be updated with minimum signaling delay and at a low signaling cost. In case there is no such link with the required capacity, a new route is found, using the optimum routing algorithm. In the FR phase, connection route is migrated to a route that has the same optimality feature with the original route. The goal of the rerouting is to establish an optimum route without applying the optimum routing algorithm after a number of handovers. This property is significant because, in the ideal case, the routing algorithm computes a single route for each connection. The optimality of the original route is maintained after the FR phase. The FHRP requires the user terminals to store information about the connection route. The performance of the FHRP is compared with a static network. In the former, the network nodes are fixed; hence there is no handover in the network. In the latter, the augmentation phase of the FHRP is applied during the intersatellite handovers; however, if a call is blocked during the path augmentation process, no rerouting attempt is made. The FHRP performs very similar to the static network and substantially better than the pure augmentation algorithm in terms of call blocking probability. Moreover, handover calls have less blocking compared to the new calls. The FHRP algorithm is applicable to connection-oriented networks.

10.7 PREDICTIVE RESERVATION POLICY

While New Adaptive Channel Reservation (NACR) uses static reservation of guard channels, the principle of Predictive Reservation Policy (PRP) consists in dynamically reserving channels when the number of communications in progress grows in a given cell. The idea is to reserve resources that will be freed when the flow is less important (when a communication ends, for example).

PRP dynamically reserves radio resources according to the local topology, because in a wireless network, traffic is really dependent on the presence of roads, homes, supermarkets, and so on. It would be very interesting to take those parameters into account while reserving channels in order to optimize the use of bandwidth. The reason is that mobiles do not move randomly: in most cases, their trajectories are foreseeable. For instance, cars follow roads and most often avoid dead-ends; only pedestrians can be found in supermarkets, and so on.

The PRP algorithm: Each cell is given a probability of transition to its neighboring cells. In a BS, when the number of occupied channels $N(t)$ reaches the threshold k or a multiple of k (Figure 10.4), the cell reserves a resource in the neighbors for which the probability of transition is high.

If they have free channels, the reservation takes place immediately. Otherwise, the PRP algorithm waits for a free channel. A reserved channel corresponds to the potential arrival

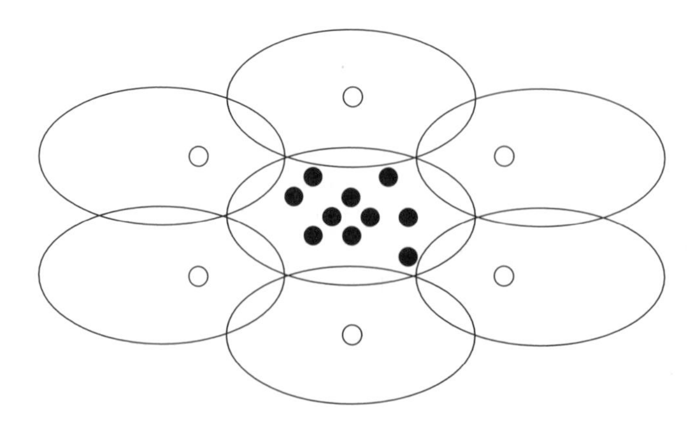

Figure 10.4 The PRP reservation principle.

of a communication. Two intervals are considered (k_0 and k_1): if the transition probability is lower than a given value p, the reservation threshold is k_0; otherwise it is k_1.

The blocked channels can only be used for handoffs. In case of overload (or a state close to the overload), the system does not accept incoming calls but only processes calls in progress. It is shown that if some cell-to-cell handoff probabilities are known (even approximately), it is possible to use them in order to manage the bandwidth efficiently. Adaptive reservation schemes prove to be efficient, especially when the traffic is unbalanced, which is a realistic assumption.

10.8 CHAINING APPROACHES

The basic idea of chaining approaches to connection rerouting is to extend the connection from the old to the new BSs, somewhat like a chain.

10.8.1 Hop-limited handoff scheme

In the PVC-based scheme, the BS in each cell is connected to the neighboring cells by a number of PVCs. If a user roams from the current cell to a new cell, the traffic path is elongated by the PVCs between the two cells. Thus, the network's processing load is not affected by the handoff frequency. Maintaining connections by continuously elongating paths from original cells to the new cells causes path inefficiency. To increase path efficiency, a hop-limited handoff scheme for ATM-based broad cellular network is used. The number of successive traffic path elongations by PVCs connecting neighboring cells is restricted to less than a predetermined number. If this number is reached, the traffic path is rerouted by the network to a PVC between the network and the new cell. Using this method, we reduce the required number of PVCs between neighboring cells and increase the traffic path efficiency. This also keeps the networking processing load light.

In a broadband network based on hierarchical ATM networks, in the planar environment, each cell is hexagonal. The BS of each cell has PVCs connected to other BSs in neighboring cells. Each BS has a number of PVCs connected to the ATM switch for the use of handoff calls only. When a mobile makes a new call, its BS will establish an SVC to carry a new call. When the call moves to an adjacent cell, the traffic path will be extended by a PVC from the current cell to the adjacent cell. In this scheme, which is called *the Hop-Limited Scheme* (HLS), a mobile is restricted to elongate its traffic path to be less than $r - 1$ times. That is, if a mobile has successfully made r handoffs, and its rth handoff request is also successful, its traffic path is rerouted from the new BS to the ATM switch. Inside the mobile, there is a counter to record the number of handoffs the mobile has made since the last path rerouting or call setup.

10.8.2 Chaining followed by make-break

Handoff also involves a sequence of events in the backbone network, including rerouting the connection and reregistering with the new AP, which causes additional loads on network traffic. Handoff has an impact on traffic matching and traffic density for individual

BSs (since the load on the air interface is transferred from one BS to another). In the case of random access techniques employed to access the air interface, or in Code-Division Multiple Access (CDMA), moving from one cell to another impacts QoS in both cells since throughput and interference depend on the number of terminals competing for the available bandwidth. In hybrid data networks, a decision on handoff has an impact on the throughput of the system. A connection management architecture is shown in Figure 10.5.

Route information is centralized in a connection server. Each ATM switch has an associated channel server, which manages local resources such as setting up translation table entries and other fabric control functions. Each BS also has an associated channel server, which manages the bridging between wired and wireless links and handling handoff requests (Figure 10.5). Connection schemes require algorithms for determining the crossover ATM switch at which rerouting occurs.

The main idea is that a real-time handoff uses the Chaining Scheme and, if necessary, a non-real-time reroute uses the make-break scheme. Ramjee *et al.* proposed and examined five different connection rerouting schemes in ATM-based wireless networks to handle handoff. They concluded that the chaining followed by make-break scheme is ideally suited for performing connection rerouting because the separation of the real-time nature of handoffs and efficient route identification in this scheme allows us to perform handoffs quickly and, at the same time, maintains efficient routes in the fixed part of the network.

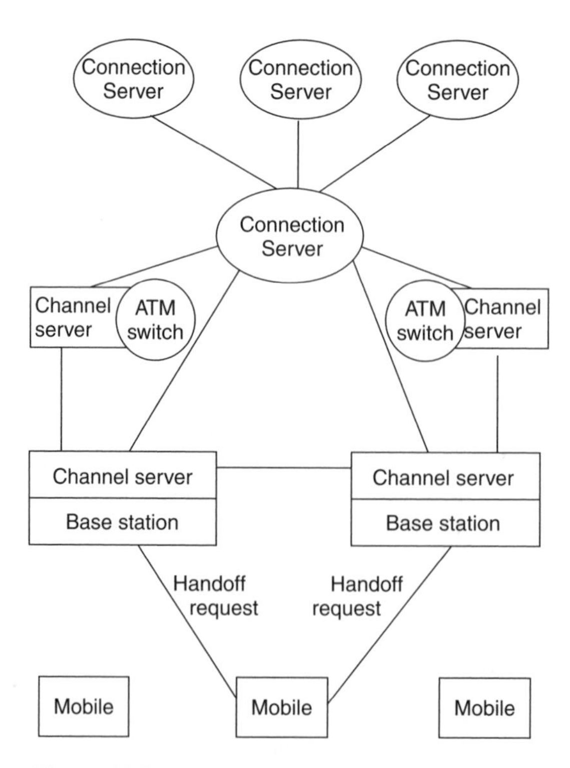

Figure 10.5 Connection management architecture.

Handoff scheme chaining followed by make-break involves the following steps:

1. The mobile host sends a handoff request message to the new BS, identifying the old BS and its connection server.
2. The new BS adds local translation table entries for its internal routing.
3. The new BS asks the old BS to forward packets pertaining to the mobile host.
4. The new BS sends back a handoff response message to the mobile host, instructing the mobile host to transmit/receive through the new station. These four steps are accomplished in real time (with a latency of 6.5 ms) to enable the mobile host to quickly switch over to the new wireless link. At this point, the chaining portion of the handoff has been completed.
5. The new BS passes the updated route information to the connection server.
6. The connection server performs necessary QoS computations on the new route. Note that it has centralized knowledge of a significant portion of the route and can perform this calculation easily. If it detects a possible QoS guarantee violation, or if fixed links are becoming congested and route efficiency is desired, it undertakes the following steps. In all other cases, the handoff flow terminates here.
7. This is the first step of the make-break portion of the handoff. The connection server identifies the best route to the COS, allocates resources along the new route, and sets up a new routing entry in the COS. The switch now multicasts cells received from the source to both the BSs.
8. The connection server informs the new BS of the completion of route change, which then starts using the new route.
9. Finally, the connection server exchanges messages with the ATM switch, tearing down the old routing entry. The connection server also requests the old and new BSs and switches in the old route to deallocate the old resources.

10.9 ANALYSIS OF CHAINING HANDOFF APPROACHES

Given a certain number of PVCs between neighborhoods and between BSs and ATM switches, call blocks in the PVC can occur. A traffic jam in the chain can degrade performance of the schemes.

If a traffic jam is unlikely, the HLS is not needed since it has more network traffic than PVC-based scheme. In such cases, the PVC-based scheme is the best selection. But with a traffic jam (high number of calls in the area), the hop-limited handoff scheme is a good selection. However, it does not consider the traffic balance in the BSs,, which belong to the ATM switch.

In chaining followed by make-break handoff scheme, the route information is centralized in a connection server. This helps to identify the COS immediately. The connection server simply determines the shortest path between the new BS and the old route, and the point of intersection is the COS. If this particular route is heavily loaded or is unable to meet the QoS requirements for the connection, the connection server will attempt to find other routes. The connection server then instructs the crossover ATM switch to reroute the connection.

In chaining followed by a make-break handoff scheme, traffic jams in the chaining part impact the performance of those schemes. The other route path, besides chaining, is calculated through the Connection Server and is optimized. This scheme only considers the traffic jams after the real-time stage during which the jams occur. Traffic jam in the chain parts should be prevented in real time.

10.10 SUMMARY

Handoff call management plays a major role in supporting acceptable levels of QoS. Connection rerouting involves limited handoff latency, maintaining an efficient route, limiting disruption of continuous media traffic, and limiting network switch update rates due to rerouting.

The approaches for connection rerouting are optimistic handoff, ordered handoff, predictive handoff, and chaining handoff.

PROBLEMS TO CHAPTER 10

Handoff in mobile and wireless networks

Learning objectives

After completing this chapter, you are able to

- demonstrate an understanding of handoff in wireless networks;
- explain different types of handoff;
- explain predictive approaches to connection rerouting;
- explain chaining followed by make-break handoff scheme.

Practice problems

10.1: How is handoff performed in wireless ATM networks?
10.2: What happens if the current QoS is not supported when handoff occurs?
10.3: What is the key idea of predictive approaches?
10.4: What is the basic idea of chaining approaches?
10.5: What is the goal of an optimistic handoff scheme?
10.6: What is the goal of an ordered approach?
10.7: What is the principle of predictive reservation policy (PRP)?
10.8: Explain PVC-based scheme.
10.9: Explain chaining followed by make-break handoff scheme.

Practice problem solutions

10.1: In WATM networks, end user devices are connected to switches *via* wired or wireless channels. The switch is responsible for establishing connections with the fixed infrastructure network component, either through a wired or a wireless channel. A mobile end user establishes a virtual circuit (VC)to communicate with another end user (either mobile or ATM end user). When the mobile end user moves from one AP to another AP, a handoff is required. To minimize the interruption of cell transport, an efficient switching of the active VCs from the old data path to the new data path is needed. Also, the switching should be fast enough to make the new VCs available to the mobile users.

10.2: When the handoff occurs, the current QoS may not be supported by the new data path. In this case, a negotiation is required to set up new QoS. Since a mobile user may be in the access range of several APs, it will select the AP that provides the best QoS.

10.3: The key idea of predictive approaches is to predict the next BS of the mobile endpoint and perform advance multicasting of data to the BS. This approach requires the maintenance of multiple connection paths to many or all the neighbors of the current BS of the mobile endpoint.

10.4: The basic idea of chaining approaches to connection rerouting is to extend the connection from the old to the new BS in the form of a chain. Chaining results in increased end-to-end delay and less efficient routing of the connection.

10.5: The goal of an optimistic handoff scheme is to perform simple and fast handoffs with the optimistic view that disruption to user traffic will be minimal. The COS simply reroutes data traffic through a different path to the new BS with the connection from the source to the COS remaining unmodified.

10.6: The goal of an ordered approach is to provide ordered lossless data delivery during handoffs. The incremental and multicast-based rerouting schemes fall into this category. However, complex protocols with resynchronization mechanisms and buffering at the BS are necessary to ensure lossless connection rerouting.

10.7: The principle of Predictive Reservation Policy (PRP) consists of dynamically reserving channels when the number of communications in progress grows in a given cell. The idea is to reserve resources that will be freed when the flow is less important (when a communication ends, for example).

PRP dynamically reserves radio resources according to the local topology, because in a wireless network, traffic is really dependent on the presence of roads, homes, supermarkets, and so on.

10.8: In the PVC-based scheme, the BS in each cell is connected to the neighboring cells by a number of PVCs. If a user roams from the current cell to a new cell, the traffic path is elongated by the PVCs between the two cells. Thus, the network's processing load is not affected by the handoff frequency. Maintaining connections by continuously elongating paths from original cells to the new cells causes path inefficiency. To increase path efficiency, a hop-limited handoff scheme for ATM-based broad cellular network is used. The number of successive traffic path elongations by PVCs connecting neighboring cells is restricted to less than a predetermined number. If this number is reached, the traffic path is rerouted by the network to a PVC between the network and the new cell. Using this method, we reduce the required number of PVCs between neighboring cells and increase the traffic path efficiency. This also keeps the networking processing load light.

10.9: In chaining followed by make-break handoff scheme, the route information is centralized in a connection server. This helps to identify the COS immediately. The connection server simply determines the shortest path between the new BS and the old route, and the point of intersection is the COS. If this particular route is heavily loaded or is unable to meet the QoS requirements for the connection, the connection server will attempt to find other routes. The connection server then instructs the crossover ATM switch to reroute the connection.

11

Signaling traffic in wireless ATM networks

Handoff algorithms in terrestrial wireless networks focus on the connection rerouting problem. Basically, there are three connection rerouting approaches: full connection establishment, partial connection reestablishment, and multicast connection reestablishment. Full connection establishment algorithms calculate a new optimum route for the call as for a new call request. The resulting route is always optimal; however, the call rerouting delay and the signaling overheads are high. To alleviate these problems, a partial connection reestablishment algorithm reestablishes certain parts of the connection route while preserving the remaining route. This way the route update process involves only local changes in the route and can be performed faster. However, the resulting route may not be optimal. In the multicast connection reestablishment algorithm, a Virtual Connection Tree (VCT) is created during the initial call admission process. The root of the tree is a fixed switching node, while the leaves are the switching centers to serve the user terminal in the future. By using the multicast connection reestablishment method, when a call moves to a cell with a new switching center, connection rerouting is done immediately owing to the already established routes. The disadvantage of this algorithm is that network resources can be underutilized as a result of resources allocated in the connection tree.

We define the Chain Routing Algorithm and implement it as a partial connection reestablishment in the handoff scheme. This process is done during chain elongation. This handoff scheme can be used in the Wireless ATM (WATM) model.

11.1 A MODEL OF WATM NETWORK

A graph $G(V, E)$ represents the topology of a WATM network. Graph G consists of two sets: a finite set V of vertices and a finite set E of edges. Graph G is represented by two subgraphs: G_1 that represents a set of ATM switching centers, and G_2 that represents a set of Base Stations (BSs). The network model is defined as follows:

- The topology of the higher level of wired subnetwork is represented by an undirected subgraph $G_1 = (V_1, E_1)$, where each edge $e \in E_1$ represents the number of communication channels and each node in V_1 represents an ATM switching center. $G_1 \subset G$, $V_1 \subset V$, and $E_1 \subset E$.
- Each edge e_i in G_1 has a limited capacity to carry a number of calls; each of these calls occupies one unit. The edges between ATM switching centers represent communication channels. The number of links is the same in each channel. The capacity of each edge is defined as $C_1(e_i)$.
- The subtopology of the BSs and their related ATM switching centers are represented by an undirected subgraph $G_2 = (V_2, E_2)$, where each edge $e \in E_2$ represents the number of channels between a base station and a switching center that are directly connected, and each node in V_2 represents a base station connected to the ATM switching center. $G_2 \subset G$, $V_2 \subset V$, and $E_2 \subset E$.
- Each edge e_i in G_2 has a limited capacity to carry a number of calls; each of these calls occupies one unit. The edges between BSs and their ATM switching centers also represent channels. The number of links in each channel is the same. The capacity of each edge is defined as $C_2(e_i)$.
- Two different BSs can establish a channel connection by allocating one edge or a sequence of edges, possibly across several ATM switching centers.
- A communication call request is denoted by $r_i = (s_1, s_2, d_1, d_2, h_1, h_2)$. This call request consists of six elements: s_1 and s_2 are the source ATM switching center and the source BS, respectively; d_1 and d_2 are the destination switching center and the destination BS, respectively; and h_1 and h_2 are the handoff switching center and the handoff BS, respectively.
- When a call request is a general call without handoff, the call request is denoted by $r_i = (s_1, s_2, d_1, d_2)$ and the handoff request options are $h_1 = 0$, and $h_2 = 0$. When a call request is a handoff request, the handoff request options are $h_1 \neq 0$ and $h_2 \neq 0$.
- For each edge e_i, which is between the ATM switching center and the BS, the total number of channels allocated for a set of call requests $R_2(r_1, r_2, \ldots, r_n)$ that arrived in the BS cannot exceed the capacity of the edge between the ATM switching center and the BS. That is, $\Sigma R_2(r_i) \leq C_2(e_i)$ for all i, $1 \leq i \leq$ number of links in a base station.
- For each edge e_i, which is among switching centers, the total number of channels allocated for a set of call requests $R_1(r_1, r_2, \ldots, r_n)$ that arrived in the switching center from the BS cannot exceed the capacity of the edge between ATM switching centers. That is, $\Sigma R_1(r_i) \leq= C_1(e_i)$ for all i, $1 \leq i \leq$ number of links between a switching center and its BSs.
- Let *Idle* $r_1(e_i)$ denote the available number of channels e_i among switching centers and *Idle* $r_2(e_i)$ denote the available number of channels e_i between switching centers and its BSs. A call request $r_i = (s_1, s_2, d_1, d_2)$ will be rejected if (*Idle* $r_1(e_i) < R_1(r_i)) \cup$ (*Idle* $r_2(e_i) < R_2(r_i)$).
- Any mobile host can access the network directly *via* a radio link to a base station that is virtually connected.

11.2 CHAIN ROUTING ALGORITHM

Handoff procedures involve a set of protocols to notify all the related entities of a particular connection for which a handoff has been executed, and the connection has to be redefined. During the process, conventional signaling and additional signaling for mobility requirements are needed. The mobile user is usually registered with a particular point of attachment. In the voice networks, an idle mobile user selects a base station that is serving the cell in which it is located. This is for the purpose of routing incoming data packets or voice calls. When the mobile user moves and executes a handoff from one point of attachment to another, the old serving point of attachment has to be informed about the change. This is called *dissociation*. The mobile user will also have to reassociate itself with the new point of access to the fixed network. Other network entities involve routing data packets to the mobile user and switching voice calls that have to be aware of the handoff in order to seamlessly continue the ongoing connection or call. Depending on whether a new connection is created before breaking the old connection, handoffs are classified into hard and seamless handoffs.

The Chaining scheme extends the connection route from the previous BS to the new BS by provisioning some bandwidth using Virtual Channel (VC) or Virtual Path (VP) reservations between neighboring BSs. Chaining can simplify the protocols and reduce signaling traffic significantly and it can be accomplished quickly. However, chaining will typically degrade the end-to-end performance of the connection and the connection route is no longer the most efficient. This could lead to dropped calls if resources in the WATM are not available for chaining. To improve the route efficiently and reduce the number of dropped calls, we propose the Chain Routing Algorithm.

We consider a broadband cellular network based on a hierarchical ATM network.

In the planar environment, each cell is hexagonal, as shown in Figure 11.1. The BS of each cell has some Permanent Virtual Circuits (PVCs) connected to the other BSs in neighboring cells. Also, each BS has a number of PVCs connected to the ATM switch

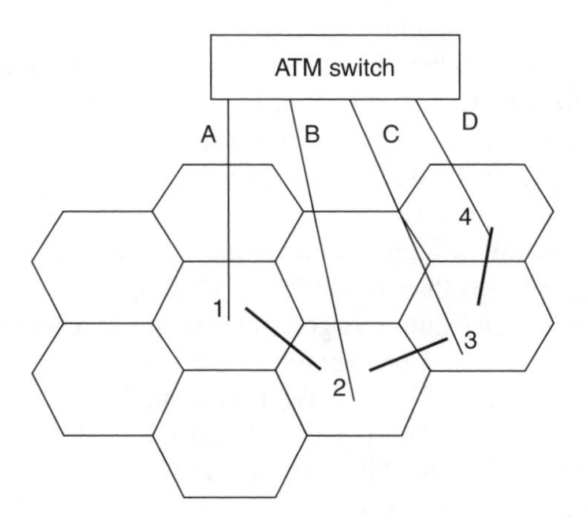

Figure 11.1 Planar personal communication network.

only for the use of handoff calls. A parameter describing Occupancy Rate of the PVC (ORP) is proposed for each BS. Overall, ORP is the larger number of occupancy rate of PVCs between the BS and its neighboring BS and the occupancy rate of PVCs between the BS and the ATM switch.

When a mobile makes a new call, its BS will establish a Switched Virtual Circuit (SVC) to carry the new call. When the terminal user moves to an adjacent cell, the traffic path will be extended by a PVC from the current cell to the adjacent cell. The chain length will be elongated by 1. Whenever the chain is elongated, one bit will be sent back to check the ORP of all the BSs on the chain route.

When the elongation is set up and all the BSs on this route have low occupancy rate, the network will follow the PVC-based scheme. In this scheme, if a user roams from its current cell to a new cell, the traffic path is elongated by the PVCs between these two cells. The traffic path will keep growing if the user keeps roaming. However, maintaining connections by continuously elongating paths from original cells to the new cells will cause the path to be inefficient.

When some parts of the route have a high occupancy rate, we propose two ways to reroute the chain parts of the route:

From the last station on the chain after each elongation, we propose sending one bit back through the chain and checking the ORP of each BS on the chain.

The path will be rerouted according to one of the following two schemes:

1. Select a route in which the length of the path is the shortest. If length of the route is shorter, it is more likely to be selected.
2. Select the path in which the PVCs have lower occupancy rate. That is, a PVC between an ATM switch and any BS in the elongation route can be set up in order to obtain a low ORP. The number of options that are available is N, where N is equal to the length of the chain.

The chain has to be rerouted whenever there is a better chain route, and the speed of elongation will be slowed down. The network efficiency can be improved significantly.

The path can be rerouted following the first scheme.

From the last station on the chain after each elongation, we send one bit back through the chain and check the ORP of each BS on the chain. If the resultant ORP of a base station is close to jam, we stop, move back one BS, and use this BS's PVC to connect to the ATM switch. If the BS at the end of the chain has a very high ORP or it is jammed, we have to send a signal to the connection server to reroute the call.

If the speed of elongation is high, the signaling and calculation cost is reduced, and the network efficiency is lower than in the Chain Routing.

We illustrate how the Chain Routing Algorithm operates by using an example. Referring to Figure 11.1, let the BS in Cell 1 be denoted as BS_1. When a mobile initiates a new call in Cell 1, BS_1 will establish an SVC between itself and the ATM switch. We consider that the mobile roams to its neighboring Cell 2 and the traffic path is elongated by the PVCs between these two cells. One bit is sent back through the chain and we check the ORP of each BS on the chain. Suppose both the BSs have low ORPs, then no rerouting occurs. We consider that the mobile roams to its neighboring Cell 3 and the traffic path is elongated by the PVCs between these two cells. One bit is sent back through the chain and the ORP of each BS on the chain is checked. Suppose both the BSs have low ORPs, then no rerouting occurs. Consider that the mobile roams to its neighboring Cell 4 and the traffic path is elongated by the PVCs between these two cells. One bit is sent back through the chain and the ORP of each BS on the chain is checked.

We have four route options:

1. Cell 4 – Link D – ATM switch
2. Cell 4 – Cell 3 – Link C – ATM switch
3. Cell 4 – Cell 3 – Cell 2 – Link B – ATM switch
4. Cell 4 – Cell 3 – Cell 2 – Cell 1 – Link A – ATM switch.

Suppose the BS of Cell 3 has a high ORP, then a new route 1 will be set up.

If part of the chain route is within one ATM switch, this chain route can be easily implemented. If the chaining route is across more than one ATM switch, this chain route method cannot be applied to the other ATM switches, because more than one ATM switch is involved, and the reroute cannot be done locally. A signal has to be sent to the connection server to reroute the call.

We illustrate how to solve this problem by using Figure 11.2.

We make a PVC neighbor link between BSs within one ATM switch and a different neighbor link connecting two BSs from two ATM switches. The neighbor link connecting two BSs from two ATM switches is a Cross ATM Switch Link (CASL). The CASL in Figure 11.2 is the link between Cell 3 and Cell 4.

The chain route has information about where it crossed more than one ATM switch. The Chain Routing Algorithm applies to the ATM area in which the chain started. When

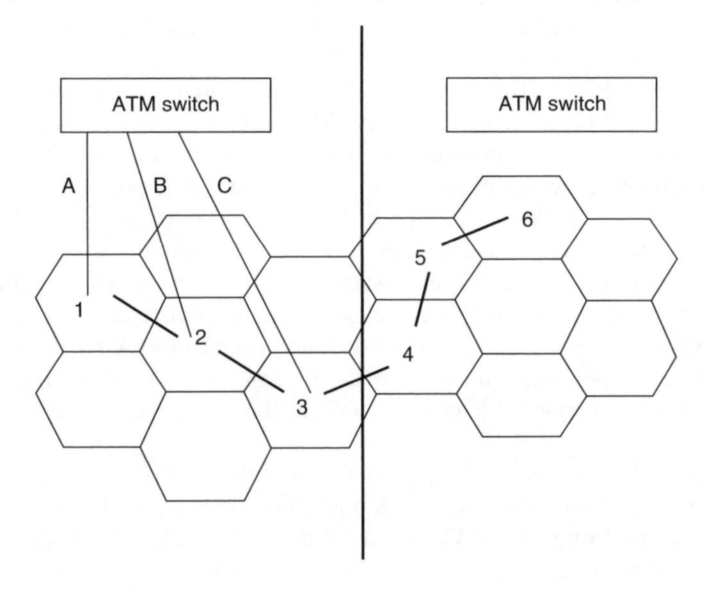

Figure 11.2 Move with more than one ATM switch involved.

the sent-back bit detects that it has arrived at the ATM switch in which the chain started, it will begin applying the Chain Routing Algorithm.

Suppose the mobile roams into Cell 6 and one bit is sent back through the chain route. When the sent-back bit sees the CASL, the Chain Routing Algorithm will be used.

In this case, we have three route options:

1. Cell 6 – Cell 5 – Cell 4 – Cell 3 – Link C – ATM switch.
2. Cell 6 – Cell 5 – Cell 4 – Cell 3 – Cell 2 – Link B – ATM switch
3. Cell 6 – Cell 5 – Cell 4 – Cell 3 – Cell 2 – Cell 1 – Link A – ATM switch.

If the BS of Cell 2 has a high ORP, for example, a new route 1 will be set up.

11.3 IMPLEMENTATION OF THE HANDOFF SCHEME

The Chain Routing Algorithm has to be implemented in the handoff scheme. The Chain Routing Algorithm is added to the handoff scheme (chaining followed by make-break) in Step 5 as follows:

1. The mobile host sends a handoff request message to the new BS identifying the old BS and its connection server.
2. The new BS adds local translation table entries for its internal routing.
3. The new BS asks the old BS to forward packets pertaining to the mobile host.
4. The new BS sends back a handoff response message to the mobile host, instructing the mobile host to transmit/receive through the new station.
5. We include the Chain Routing Algorithm. A single bit is transferred from the mobile host back to the starting point of the chain route. It checks the ORP of each BS. After the new route is found and the new BS chosen, which is connected to the ATM switch, the new BS sends a message to the ATM switch channel server (performing make, break, and break-make). The new BS can change its translation table entries in its BS channel server immediately and the new connection between the chain to the ATM switch is established. This way, the chaining portion of the handoff is completed. Note that these five steps 1, 2, 3, 4, and 5 are accomplished in real time.
6. The new BS passes the updated route information to the connection server.
7. The connection server performs necessary Quality-of-Service (QoS) computations on the new route. Note that the connection server has centralized knowledge of a significant portion of the route and can perform this calculation easily. If the connection server detects a possible QoS guarantee violation, or if the fixed links are becoming congested and route efficiency is desired, the connection server undertakes the following steps: 8, 9, and 10.
   In all other cases, the handoff flow terminates at this point.
8. This is the first step of the make-break portion of the handoff. The connection server identifies the best route to the Crossover Switch (COS), allocates resources along the new route, and sets up a new routing entry in the COS. The switch multicasts cells received from the source to both BSs.

9. The connection server informs the new BS of the completion of the route change, which then starts using the new route.
10. The connection server exchanges messages with the ATM switch, removing the old routing entry. The connection server also requests the old and new BSs and switches in the old route to release the old resources.

11.4 ANALYSIS OF THE CHAIN ROUTING ALGORITHM

Upon receiving a handoff request from the mobile host, the new BS first executes the procedures for the Chain Routing Algorithm scheme. The new BS then transmits the handoff response message to the mobile host so that the mobile host starts listening and transmitting *via* the new BS. The new BS then initiates the make-break rerouting procedure. The scheme combines the advantages of both make-break and Chaining schemes. It results in fast handoffs so that the mobile host is quickly connected to the new BS during handoff. Furthermore, an optimistic scheme can be later employed, as needed, in order to make more effective use of bandwidth and to minimize disruption. This scheme is useful in cases when a user is handed over in a network that is lightly loaded or when the mobile user does not travel far during a connection. In such cases, the handoff performed using chaining does not disrupt the communication, and since the network is lightly loaded, there will be no noticeable performance degradation due to the increased hop count. If the network becomes congested or if the user moves far enough so that the effects of the extended chain are undesirable, the make-break scheme can be applied to reroute the connection.

11.4.1 Comparison of chain routing algorithm with Hop-limited method

The elongation pattern in the Chain Routing Algorithm is one adjustment of the Hop-limited handoff scheme and it is based on the Chaining scheme. By analyzing the Chaining scheme and the Hop-limited handoff scheme, we compare the results with Chain Routing Algorithm scheme. Akyildiz *et al.* present performance analysis of the Hop-limited handoff scheme and Chaining scheme. We make the following assumptions.

1. The call holding time T_M is exponentially distributed with mean $1/\mu_M$.
2. The originating calls arrive in a cell following a Poisson process with rate λ_o.
3. The time interval R during which a mobile resides in a cell called the *cell sojourn time* has a general distribution. The cell sojourn times, $R^{(1)}, R^{(2)}, \ldots$, are independent and identically distributed.

We consider a mobile in a cell. A Virtual Circuit (VC) connecting the cell's BS to the ATM switch or to an adjacent cell's BS is occupied by the mobile. The VC can be released in three cases: (i) the connection is naturally terminated; (ii) the connection is forced to be terminated due to handoff blocking; and (iii) the mobile has already successively made $r - 1$ handoffs since it came to the current cell and it is making the rth handoff attempt (here r is a system parameter). Let the time interval from the moment the VC is occupied by the call to the moment the VC is released be T_r and we will derive the VC's holding time.

First we only consider cases (ii) and (iii). Let P_f be the probability that the call is blocked due to unavailability of the PVC when the mobile tries to handoff to another cell. Let θ_r be the VC's holding time under this consideration and n be the number of handoffs the mobile will try from the moment it comes to the cell to the moment the VC is released. For $1 \leq i < r$, $p(n = i) = p_f(1 - p_f)^{i-1}$; $p(n = r) = (1 - p_f)^{r-1}$.

Let $N[z]$ be the generating function of variable n, and $\theta_r^*(s)$ be the Laplace–Stieltjes Transform (LST) of θ_r. We have $\theta_r^*(s) = N[R^*(s)]$, where $R^*(s)$ is the LST of the distribution function of R. The distribution of T_r is $F_{Tr}(t) = Pr(\min(\theta_r, TM) \leq t)$. For the assumption that R is also exponentially distributed with mean μ_R, the mean of T_r is $E[T_r] = \{1 - [\mu_r(1 - P_f)/(\mu_M + \mu_R)]^r\}/(\mu_M + P_f\mu_R)$.

Let us derive the handoff call arrival rate. There are two kinds of handoff calls. The first type of handoff call will request a PVC connecting the BS to the ATM switch, with mean arrival rate λ_{h1}; the handoff call of another type will request a PVC connected to its previous cell's BS, with mean arrival rate λ_{h2}. Let p_i be the probability that a call will make the ith handoff request. Then we have $p_i = (1 - p_f)^{i-1}[\mu_R/(\mu_M + \mu_R)]^i$.

Assume the probabilities of a handoff call coming from arbitrary neighboring cells are the same. Let N_1 be the mean number of SVCs connecting a cell to the ATM switch. We have $N_1 = \lambda_o(1 - p_n)E[T_r]$.

Let N_2 be the number of required PVCs connecting each BS to the ATM switch for rerouting requests. We can model this as an $M/M/m/m$ queuing system, where the arrival rate is λ_{h1} and the average holding time is $E[T_r]$. Thus, we have

$$P_f = \frac{[((\lambda_{h1}E[T_r])^{N_2}/N_2!)]}{\sum_{n=2}^{N_2}(\lambda_{h1}E[T_r])^n/n!}$$

Let N_3 be the required PVCs to connect the BS to a neighboring BS. This example is more complex, and we calculate the upper bound. Assume the mean holding time of all PVCs is $E[T_{r-1}]$, then we can model this case as an $M/M/m/m$ system, with six neighboring cells, the arrival rate is $\lambda_{h2}/6$ and mean holding time $E[T_{r-1}]$. We have

$$P_f = \frac{((\lambda_{h2}E[T_{r-1}]/6)^{N_3}/N_3!)}{\sum_{n=0}^{N_3}(\lambda_{h2}E[T_{r-1}]/6)^n/n!}$$

Using this equation we can obtain N_3 to satisfy the P_f requirement.

Now we roughly compare the Hop-Limiting Scheme (HLS) with the VCT and Chaining scheme. Assume there are 49 cells. Table 11.1 shows the required number of VCs for different schemes, given the new call arrival rate is 11.9 calls per minute, the mean call holding time is 2 min. and the mean call sojourn time is also 2 min. The new call blocking probability is 0.01 and handoff call blocking probability is 0.001. The row $r = \infty$ shows

Table 11.1 Required number of VCs for different schemes

	N_1	N_2	N_3	Total
$r = 1$	11.78	24	0	35.78
$r = 3$	20.6	10	54	84.6
$r = 5$	22.8	5	66	93.8
$r = \infty$	23.56	0	72	97.56
VCT	1155	0	0	1155

the requirements of the Chaining scheme. When r is finite, the number of required VCs is lower than those of the Chaining and VCT schemes. This means that the bandwidth efficiency is higher. When $r = 1$, the number of required VCs is the smallest, but during each handoff a network is evoked to reroute the traffic path, which means that the network processing load is the heaviest. When choosing a value for r, there is a trade-off between the number of required VCs and the ATM switch processing load.

Comparing the Hop-limited handoff scheme with relatively big r values, the Chain Routing Algorithm tends to use higher number of required PVCs (N_2) connecting each BS to the ATM switch for rerouting requests. When the occupancy rate of the route path increases, the Chain Routing Algorithm needs to revoke more rerouting at the chain part of the route. At the same time, the Chain Routing Algorithm tends to use lower number of PVCs (N_3) to connect the BS to the neighboring BS compared with the Hop-limited handoff scheme with relatively small r values. Because the Chain Routing Algorithm needs to revoke more rerouting at the chain part of the route, generally the length of the route is smaller than in the Hop-limited handoff scheme with relatively small r values.

Chain Routing Algorithm produces less signaling traffic and network processing load than the Hop-limited handoff scheme with a small number of r, because it will not evoke the network to reroute the traffic path so often. At the same time, it has lower bandwidth efficiency than the Hop-limited handoff scheme with small number of r, because it will need more VCs to connect the BS to a neighboring BS.

Chain Routing Algorithm produces more signaling traffic and network processing load than the Hop-limited handoff scheme with a large number of r, because it needs to do a rerouting process in the chaining parts and it evokes the network to reroute the traffic path more often. At the same time, it has higher bandwidth efficiency than the Hop-limited handoff scheme with large number of r, because it will need less VCs to connect the BS to a neighboring BS.

The Chain Routing Algorithm is another option that can be selected besides the Hop-limited handoff scheme. It can give better performance than the Hop-limited handoff scheme in certain cases. Its performance can be adjusted by tuning the threshold at which it performs the chain routing calculation.

11.4.2 Analysis of the signaling traffic cost

Signaling traffic is caused by reroute-related updates and modifications occurring in the ATM switches. In the Chaining scheme we can provision bandwidth between neighboring

BSs and thereby avoid modifying the switch routing entries. Thus, there is a clear trade-off between the amount of bandwidth provisioned and the number of reroute updates. The amount of provisioned bandwidth can be used as a tunable parameter for engineering network resources.

We analyze the signaling traffic cost in the Chain Routing Algorithm scheme. When no reroute is found, the starting BS has a bit that remembers this BS is the starting point of the chain. The signal needs to be transferred in a single bit that is transferred from the mobile host back to the starting point of the chain when the mobile host performs a handoff.

When rerouting is needed, one message is sent to the ATM switch channel server and one message is sent to the BS server. The messages to the ATM switch channel server contain the necessary 3-tuple [Virtual Path Identifier (VPI), Virtual Channel Identifier (VCI), and port] for modifying the switch translation table entry. The messages to the BS channel server (add entry, delete entry, delete forwarding entry, and forward) also contain only the necessary 3-tuples for the BS to update its translation table entries.

Because QoS computation is not involved, the Chain Routing Algorithm scheme can be performed in real time. It reduces the risk of a lost connection because of limitation of bandwidth availability and it improves the efficiency of the PVC between neighboring BSs and between the BS and the ATM switches. A possible QoS guarantee violation or congested fixed links are reduced because previous routes before handoff are optimized through connection server. The most likely problem is the handoff part. If the chain part is improved, the entire route is improved. As a result, the chance of going through Steps 8 and 9 and the signaling traffic involved in 8 and 9 is reduced.

Signaling traffic depends on the network configuration and protocols involved. In the simulation model, when a mobile user roams within the ATM switch area, the signaling traffic is low in the Chain Routing Algorithm scheme. It performs like the Chain Routing Algorithm scheme. When a rerouting process is required, the signaling messages are a few bytes long because only one ATM switch is involved. The longest message is the handoff request message from the mobile user. This message is 44 bytes long and includes the mobile identity, old BS channel server identifier, and the 3-tuple (VPI, VCI, and port) of the translation table entry at the same ATM switch. The route update message to the connection server contains the identity of the mobile endpoint and the two BSs involved in the Chaining scheme.

When the mobile user roams outside the original ATM switch and a reroute is requested because of the overload of links or QoS problem, the new BS needs to identify the best route to the COS, allocate resources along the route, and then exchange messages with the COS, which executes break-make or make-break operations. In this case, the Chain Routing Algorithm performs better than the Chaining scheme.

1. In certain cases, because of the Chain Routing Algorithm, the links connecting BSs and the links connecting the ATM switch and the BS are utilized more efficiently, so this kind of reroute does not occur as often as in the Chaining scheme.
2. In certain cases, when the mobile user roams to the other ATM switch area, the chain part inside the original ATM switch area will be rerouted according to the Chain Routing Algorithm, so this portion of the routing path will probably not have overload

problems and QoS problems as the Chaining scheme does, and the overall routing path is not likely to be rerouted as in the Chaining scheme.

The difference can be demonstrated by the different call-drop rates in certain network configurations.

11.4.3 Handoff latency

The Chaining scheme and Chaining with Break-Make and Make-Break extends the connection route from the previous BS to the new BS. By provisioning some bandwidth by using virtual channel (VC) reservations between neighboring BSs, the chaining can be accomplished quickly (since the COS is not involved). However, chaining will typically degrade the end-to-end performance (e.g., end-to-end delay) of the connection, and the connection route is no longer the most efficient. This can lead to dropped calls if resources in the wired network are not available for chaining.

Handoff latency is defined to be the time duration between the following two events at the mobile host: the initiation of handoff request and the reception of handoff response. Table 11.2 lists the handoff latencies incurred by the five connection rerouting schemes.

The handoff latency is slightly higher for the break-make scheme as compared to the make-break scheme because the break-make scheme involves two operations (break and make) at the switch before the handoff response can be sent, whereas only one operation (make) is needed in the make-break scheme. The Chaining scheme is fast because it preassigns VCs between neighboring BSs and, thus, translation entries at the COS need not be changed. If VC's were not preassigned, the handoff latency in the Chaining scheme would be comparable to that of the make-break scheme.

Chaining with break-make and Chaining with make-break perform their rerouting operations after handoff and, thus, those operations do not affect the handoff latency of these schemes. Also, note that the handoff latency measurements depend on the number of connections of the mobile endpoint that must be rerouted. This is because each connection corresponds to a translation table entry in the switch. Therefore, rerouting multiple connections implies that multiple translation table entries have to be modified, resulting in higher latencies. Regarding the impact of connection rerouting involving multiple ATM switches on handoff latency, the handoff latency in Chaining with break-make and Chaining with make-break will not be affected since latency is determined only by the chaining

Table 11.2 Handoff latency in connection rerouting schemes

	Latency for 1 connection (ms)
Rerouting scheme	46.4
Make-break (m-b)	37.7
Chaining	6.5
Chaining with (b-m)	6.5
Chaining with (m-b)	6.5

of the neighboring BSs. On the other hand, in break-make and make-break schemes, handoff latency is directly proportional to the number of ATM switches that need to be updated along the new route. Thus, the separation of connection rerouting from the real-time phase in Chaining with break-make and Chaining with make-break schemes, results in low handoff latency regardless of the number of switches involved in the rerouting operations. In the Chain Routing Algorithm scheme, the Chain Routing Algorithm only applies to the chain part of the route path, translation entries at the COS need not be changed. The time cost of chain routing attributed to handoff latency will be comparable to the Chaining Algorithm.

In certain cases, some calls are blocked because of the overload of the chain part of the route path and those links between the ATM switch and the BSs. In cases in which those calls have been rerouted, the handoff latency is comparable to the make-break scheme or the break-make scheme. For the HLS and Chain Routing Algorithm scheme, the routes have to be rerouted in certain circumstances.

For HLS, when a mobile has successfully made $r - 1$ handoffs, and its rth handoff request is also successful, its traffic path would be rerouted from the new BS to the ATM switch to which it belongs. Regardless of whether the mobile user roams out of the current ATM switch area to a new ATM switch or not, the connection server performs make-break or break-make and necessary QoS computations on the new route. If the mobile user roams out of the current ATM switch area, the handoff latency is directly proportional to the number of ATM switches that need to be updated along the new route.

The handoff latency is similar to a make-break or break-make scheme. For the Chain Routing Algorithm scheme, the route will be rerouted when the occupancy of the VCs between the current BS and its neighboring BS or of the VCs between an ATM switch and the base station reach a certain value. Two cases are considered: one when the user roams inside an ATM switch area and the other when the user roams outside the current ATM switch area and the handoff latency is different from HLS. Regardless of whether the user roams inside an ATM switch area or the user roams outside the current ATM switch area, only the chain part of the route path inside the original ATM switch area will be rerouted. The handoff latency is similar to the Chaining scheme. Because handoff latency of the Chain Routing Algorithm consists of rerouting cost and chaining cost and handoff latency of the Chaining scheme consists of chaining cost only, the latency of the Chain Routing Algorithm is higher than that of the Chaining scheme. Depending on the r value of HLS, the latency of Chain Routing Algorithm is higher than that of the HLS with a large r value but less than that of an HLS with a small r value. Suppose those routes that are blocked need to be rerouted as those in break-make and make-break schemes. The average handoff latency of different schemes can be estimated as follows:

1. Chaining latency (during elongation, estimated to 6.5 ms).
2. Chain Routing Algorithm scheme latency (during elongation, estimated to 6.5 ms).
3. Rerouting cost (rerouting inside one ATM switch, estimated to 6.5 ms).
4. Rerouting cost (rerouting outside one ATM switch, estimated to 45 ms).

Regarding the impact of connection rerouting involving multiple ATM switches on handoff latency, the handoff latency in Chaining with make-break will not be affected since latency is determined only by the chaining of the neighboring BSs. Thus, the separation of connection rerouting from the real-time phase in Chaining with make-break results in a low handoff latency regardless of the number of switches involved in the rerouting operation. While connection rerouting due to handoffs is similar to rerouting due to the failure of network components, there are two important differences. First, handoffs are much more frequent than network faults. With frequent reroutes, the disruption caused to ongoing connections has to be minimized. On the other hand, in many cases applications will be willing to tolerate some disruption due to rare network fault rerouting scenarios. Second, handoffs result in connection reroutes that are limited to a small geographic locality (e.g., neighboring BSs). On the other hand, reroutes due to failures may involve reestablishing the entire connection.

In ATM networks, all data is transmitted in small, fixed-size packets. Owing to the high-speed transfer rate (in the range of hundreds to thousands of $\mathrm{Mb\,s^{-1}}$) and rather short cell length (53 bytes), the ratio of propagation delay to cell transmission time and the ratio of processing time to cell transmission time of ATM networks will increase significantly more than that in the existing networks. This leads to a shift in the network's performance bottleneck from channel transmission speed (in most existing networks) to the propagation delay of the channel and the processing speed at the network switching nodes. This chain routing method will decrease the workload in the network switching nodes.

1. There is no need to identify the COS when rerouting, because chain routing method works only with one ATM switch.
2. There is no need to calculate the best route through the connection server because it is done locally to reduce signaling traffic.
3. It is easy to implement. Only one parameter ORP is added to the new scheme and the calculation is very simple. It complies with existing ATM signaling standards and its implementation leaves commercially available ATM components unaffected.
4. It is in real time.
5. It can significantly reduce signaling traffic.
6. In the new handoff scheme, the concern of traffic jam is included. This scheme can handle different kinds of situations efficiently. By doing this, the whole PVC in this ATM switch will have the highest utility efficiency, so that the system adopting this scheme can handle many more handoffs.

The Code of the Chain Routing Algorithm is as follows:
Each cell has the following parameters:

- ORP1 – overall occupancy rate of PVCs to ATM switch
- ORP2 – overall occupancy rate of PVCs between neighbors (in one direction)
- Chain length – Number of BSs on the chain after CASL.
- ROU – new route needs to be implemented
- CASL – Cross ATM Switch Link.

The ALGORITHM is as follows:

```
CASL=NULL;
ROU=NULL;
```

```
If mobile roams to a new BS
       If the link = CASL
                CASL !=NULL
Else Chain length ++;
Traffic path will be extended by a PVC from the current cell to
the adjacent cell
While (CASL !=NULL) on the chain route
      Go through the route until pass CASL
For (i=0, i<Chain Length, i++)
Route [i]=From the end of the chain (from which BS the mobile has
just arrived)
go through i number of BSs and go to the ATM switch.
For (j=0, j<i, j++)
Check the ORPs of the BSs on Route[i]
If (ORP1=jammed) or (ORP2=jammed)
Determine from Route[0-j], which Route has the lowest ORP
            Record the Route as ROU
Terminate the loop.
            End IF
ORP of Route[j]=highest ORP of those BSs
If ROU !=NULL
Reroute the chain route as ROU
```

11.5 SUMMARY

The handoff call management scheme reduces signaling traffic in the wireless ATM network and improves the efficiency of virtual channels. Chaining followed by make-break algorithm is a suitable handoff scheme for various situations. In the chaining part of the scheme, a chain routing algorithm is studied and compared with the HLS. The implemented algorithm improves the performance of the existing scheme in call-drop rates so as to reduce the signaling traffic in the WATM network. This method complies with the existing ATM signaling standard and its implementation leaves commercially ATM components unaffected. It considers the traffic condition when chaining and it is easy to implement in the chaining followed by the make-break scheme.

PROBLEMS TO CHAPTER 11

Signaling traffic in wireless ATM networks

Learning objectives

After completing this chapter, you are able to

- demonstrate an understanding of connection rerouting;
- explain the role of a chaining scheme;
- explain how signaling traffic occurs;
- explain handoff latency;

Practice problems

11.1: What are the connection rerouting approaches?
11.2: What are the schemes for path rerouting?
11.3: How is chain routing implemented in the handoff scheme?
11.4: What does the new base station (BS) do upon receiving a handoff request?
11.5: What is the cause for the signaling traffic?
11.6: What is handoff latency?

Practice problem solutions

11.1: There are three connection rerouting approaches: full connection establishment, partial connection reestablishment, and multicast connection reestablishment.

11.2: The path is rerouted according to the shortest path scheme, or the path in which the PVCs have lower occupancy rate is selected.

The chain has to be rerouted whenever there is a better chain route and the speed of elongation will be slowed down. The network efficiency can be improved significantly.

11.3: Implementing chain routing involves transferring of a single bit from the mobile host back to the starting point of the chain route. It checks ORP of each BS. After the new route is found and the new BS, which is connected to the ATM switch is chosen, the new BS sends a message to the ATM switch channel server (performing make, break, and break-make). The new BS can change its translation table entries in its channel server immediately and the new connection between the chain to the ATM switch is established.

11.4: Upon receiving a handoff request from the mobile host, the new BS first executes the procedures for the Chain Routing Algorithm scheme. The new BS then transmits the handoff response message to the mobile host so that the mobile host starts listening and transmitting *via* the new BS. The new BS then initiates the make-break rerouting procedure.

11.5: Signaling traffic is caused by reroute-related updates and modifications occurring in the ATM switches.

11.6: Handoff latency is defined as the time duration between the following two events at the mobile host: the initiation of handoff request and the reception of handoff response.

12

Two-phase combined QoS-based handoff scheme

Wireless Personal Communication Services (PCS) and broadband networking for delivering multimedia information represent two well-established trends in telecommunications. While technologies for PCS and broadband communications have historically been developed independently, harmonization into a single architectural framework is motivated by an emerging need to extend multimedia services to portable terminals. With the growing acceptance of Asynchronous Transfer Mode (ATM) as the standard for broadband networking, it has become appropriate to consider the feasibility of standard ATM services into next-generation microcellular wireless and PCS scenarios. The use of ATM protocols in both fixed and wireless networks promises the important benefit of seamless multimedia services with end-to-end Quality-of-Service (QoS) control. The wireless ATM (WATM) specification provides an option to existing ATM networks that wish to support terminal mobility and radio access while still retaining backward compatibility with ATM equipments.

The current developments on WATM are mainly based on ATM as the backbone network with a wireless last-hop extension to the mobile host. Mobility functions are implemented into the ATM switches and the Base Stations (BSs). WATM helps to bring multimedia to mobile computers. Compared with the wireless LANs, which have a limitation of bandwidth to support multimedia traffic and slow handoff, the bandwidth of existing mobile phone systems is sufficient for data and voice, but it is still insufficient for real-time multimedia traffic. ATM has more efficient networking technology for integrating services, flexible bandwidth allocation, and service type selection for a range of applications. The current interest and research efforts are intense enough to claim that WATM will continue to be pursued as a research and development topic in the next few years.

There are two major components in WATM networks:

1. A radio access layer providing high-bandwidth wireless transmission with appropriate Medium Access Control (MAC), Data Link Control (DLC), and so on.

2. A mobile ATM network for interconnection of BSs [Access Points (APs)] with appropriate support of mobility related functions, such as handoff and location management.

We focus on the mobile ATM handoff control required to support Mobile Terminal (MT) migration from one WATM microcell BS to another. The handoff function should ensure that the ongoing connection is rerouted to another AP in a seamless manner. The design goal of the handoff in WATM is to prevent service disruption and degradation during and after the handoff process.

To support wireless users in an ATM network, the main challenges are due to the mobility of the wireless users. If a wireless user moves while communicating with another user or a server in the network, the network may need to transfer the radio link of the user between radio APs in order to provide seamless connectivity to the user. The transfer of a user's radio link is referred to as handoff. During a handoff event, the user's existing connection may need to be rerouted in order to meet delay, QoS or cost criteria, or simply to maintain connectivity between two users, or between a server and wireless users. Rerouting is critical to wireless networks that need to maintain connectivity to a wireless user through multiple, geographically dispersed radio APs. Rerouting must be done quickly to maintain connectivity to the network during a handoff event. In addition, the resulting routes must be optimum. A two-phase interswitch handoff scheme meets the requirement of the rerouting. In the first phase, connections are rapidly rerouted and in the second phase a route optimization procedure is executed. For the two-phase handoff scheme, the first phase is simply implemented by path extension and the second phase is implemented by partial path reestablishment.

We describe the QoS-based rerouting algorithm that is designed to implement two-phase interswitch handoff scheme for WATM networks. We use path extension for each interswitch handoff, and invoke path optimization when the handoff path exceeds the delay constraint or maximum path extension hops constraint. We study three types of path optimization schemes: combined QoS-based, delay-based and hop-based path rerouting schemes.

We use QoS combined path optimization scheme for WATM network. We focus on the problems related to the support of mobility in the WATM network. This scheme determines when to trigger path optimization for the two-phase handoff and how to minimize the service disruption during path optimization.

12.1 WIRELESS ATM ARCHITECTURE

A WATM network is intended to support integrated broadband services to MTs through an ATM User Network Interface (UNI). Figure 12.1 shows a network diagram that illustrates various network entities and the functions that are required to support mobility in such an ATM network.

In this architecture, the MT is an ATM end system that can support multimedia applications. The wireless link between the MT and BS provides the desired ATM transport services to the MT. A mobility-enhanced signaling protocol based on the ITU recommendation Q.2931 is used by the MT, BS, and Mobility Support Switches (MSS) to support handoff-related functions.

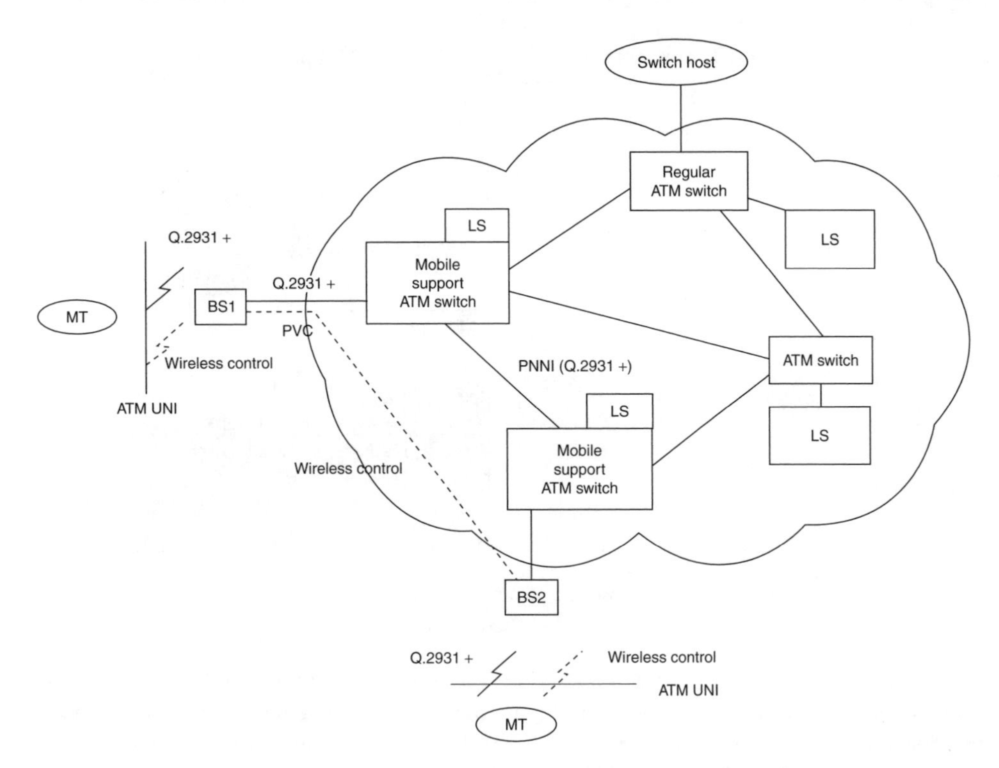

Figure 12.1 Network entities and functions for mobility support in ATM network.

There are two types of interfaces in ATM: the UNI and Network-to-Network Interface (NNI). Both interfaces can be private or public depending on the network. In the location management scheme that is proposed as Location Server (LS), a database is required to register the location of the MT.

The ATM Forum designed the WATM specifications, which are compatible with standard ATM protocols by providing ATM-based radio access as well as extensions for mobility support within an ATM network. The wireline and wireless ATM protocol stacks are shown in Figure 12.2, in which the shaded areas represent new sublayers added for wireless support.

At the bottom of the protocol stack is the physical layer. In order to support a wide range of multimedia applications, the WATM physical layer should provide reasonably high data rates. A MAC protocol in WATM is needed to meet the following requirements:

- It should be able to work with the upper-layer protocol seamlessly.
- The MAC layer should be designed to use bandwidth efficiently to accommodate a reasonably large number of users.
- The MAC protocol should guarantee a certain QoS to the user for various services, such as Constant Bit Rate (CBR), Variable Bit Rate (VBR), Available Bit Rate (ABR), and Unspecified Bit Rate (UBR).

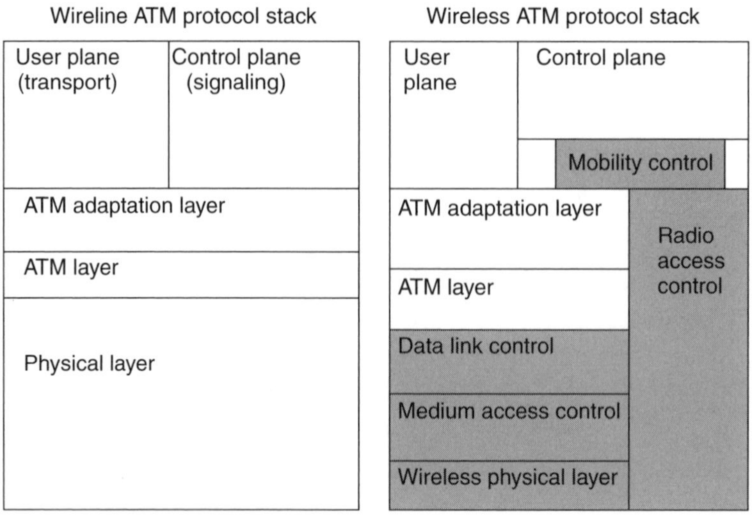

Figure 12.2 Wireline and wireless ATM protocol stacks.

The DLC protocol can provide error control by retransmitting damaged or lost frames. To prevent a fast sender from overrunning a slow receiver, the data link protocol can also provide flow control. The sliding window mechanism is widely used to integrate error control and flow control.

The purpose of the ATM Application layer is to convert the data from a higher layer (e.g., the application layer) into a format that is suitable for transmission over ATM cells. In other words, the ATM Adaptation-Layer (AAL) protocol provides an interface between the application layer and the ATM network layer. The AAL protocol is available to adapt the different applications without sacrificing its inherent advantages – low delay and fast transport. The AAL is broken down into two sublayers: the Convergence Sublayer (CS) that performs service-dependent functions and the Segmentation and Reassembly (SAR) sublayer that performs segmentation and reassembly. In order to define the functions of the CS, we can divide the services into four classes as shown in Table 12.1.

On the basis of the different types of services as shown in Table 12.1, the adaptation-layer protocol has been classified into five types. AAL1 (voice) emulates a synchronous, CBR connection. AAL2 (video) is suitable for the traffic that has a bit rate that varies in time and requires delay bounds such as compressed video. AAL3 and AAL4 as well as AAL5 (packet transfer) provide frame segmentation and reassembly functions and are suited for variable traffic without delay requirement. The ATM Forum agreed that AAL1 will support CBR service and AAL5 will support all other services. By far, the AAL5 is the most important of the AALs. AAL5 connections allow ATM networks to interface with the Internet's transport protocol, TCP/IP, by packaging the IP packets into ATM cells.

Table 12.1 Types of services and attributes

Service class	Attributes
A	Connection oriented, CBR, needs to transmit timing information over the ATM cells, e.g., circuit emulation.
B	Connection oriented, VBR-RT, needs to transmit timing information over the ATM cells, e.g., multimedia service with VBR video and audio.
C	Connection oriented, VBR-NRT, ABR, UBR, does not need to transmit timing information over the ATM cells, e.g., traditional data traffic such as X.25.
D	Connectionless, VBR-NRT, ABR, UBR, does not need to transmit timing information over the ATM cells, e.g., e-mail service.

The mobility control sublayer immediately above the MAC layer performs control functions related to the physical radio channel control and metasignaling between the MT and BSs (e.g., terminal initialization, handoff, and power control).

12.2 MOBILITY SUPPORT IN WIRELESS ATM

A key feature of any wireless network is the capability to support handoff. Handoff is an action of switching a call in progress in order to maintain continuity and the required QoS of the call when a MT moves from one cell to another. In a mobile ATM network, an MT can have several active links with different QoS requirements. These Virtual Channels (VCs) with different QoS introduce challenges to the handoff protocol. In general, the handoff with multirate ATM connections must be supported with low cell loss, latency, and control overhead. The QoS constraints for each individual connection should be maintained during the MT migration.

There are several types of handoff. We can classify the types of handoff on the basis of the number of active connections and the direction of the handoff signaling. We describe these types of handoff as follows:

On the basis of the number of active connections

The handoffs can be classified on the basis of the number of connections that an MT maintains during the handoff procedure. There are two types of handoffs based on this classification: hard handoff and soft handoff.

In hard handoff, the MT switches the communication from the old link to the new link. Thus, there is only one active connection from the MT at any time. There is a short interruption in the transmission. This interruption should be minimized in order to make the handoff seamless.

In soft handoff, the MT is connected simultaneously to two APs. As it moves from one cell to another, it 'softly' switches from one BS to another. When connected to two BSs, the network combines information received from two different routes to obtain a better quality. This is commonly referred to as macrodiversity.

On the basis of the direction of the handoff signaling

Another way of classifying the handoff is on the basis of the direction of the handoff signaling. There are two types of handoffs based on this classification: forward handoff and backward handoff.

In forward handoff, after the MT decides the cell to which it will make a handoff, it contacts the BS controlling the cell. The new BS initiates the handoff signaling to link the MT from the old BS. This is especially useful if the MT suddenly loses contact with the current BS.

In backward handoff, after the MT decides the cell to which it attempts to make a handoff, it contacts the current BS, which initiates the signaling to handoff to the new BS.

There are five types of handoff schemes: handoff using full reestablishment, handoff using multicasting, handoff using connection extension, handoff using partial reestablishment, and handoff using two-phase protocol.

Handoff using full reestablishment

In a connection-oriented wireless environment, virtual circuits are established from the source to the destination. The data follows the path that has been set up, and an in-order delivery is guaranteed. If a handoff is to occur, the old virtual connection is torn down and an entirely new virtual circuit is set up from the current source to the current destination. Since both ends are explicitly involved, this handoff scheme is not transparent. Severe traffic interruptions are experienced and hence this scheme is not recommended. Figure 12.3 shows the handoff using full reestablishment.

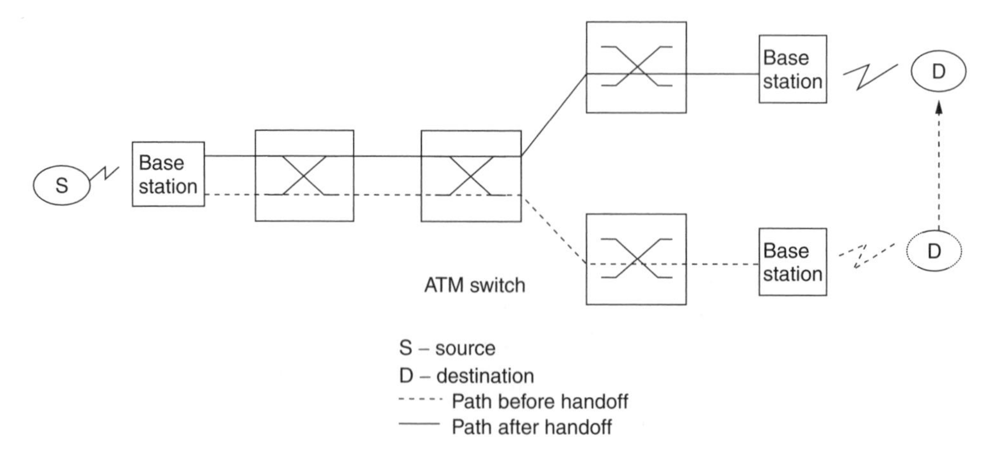

Figure 12.3 Handoff using full reestablishment.

Handoff using multicasting

Multicasting is used to support handoffs in both the connection oriented and connectionless scenarios. In the case of WATM environment, multicasting is used to establish links to all BSs that are neighboring the BS that is currently controlling a MT. Subsequently, in whichever direction the MT moves, a handoff path has already been established. Also, since the data is being multicast, it continues to flow without any interruption. This scheme ensures a lossless and seamless handoff. However, since data is being multicast to the entire set of nodes, most of which is unused, bandwidth is being utilized very inefficiently. Also, if an MT is at the edge of two cells, it is very likely that it might get two copies of the data packets. This leads to other complications like BS synchronization. Thus, this scheme is not recommended. Figure 12.4 shows the handoff using multicasting.

Handoff using connection extension

The basic idea of this scheme is that the local paths are more affordable than the global paths. When an MT migrates from one BS to another, the old BS extends the connection to the new BS. The obvious disadvantage of this method is that the new path to the MT is not an optimal path. Figure 12.5 shows the handoff using the connection extension scheme.

Handoff using partial reestablishment

This scheme is certainly better than reestablishing a new connection from the source to the destination or extending an existing connection to the new BS. This scheme uses the concept of a Crossover Switch (COS). The new BS does a partial reestablishment of the

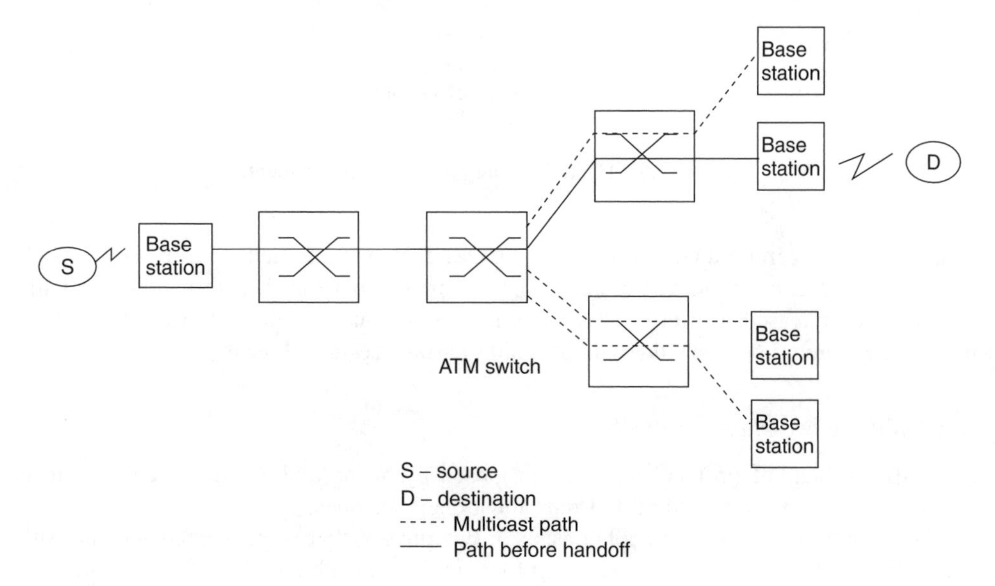

Figure 12.4 Handoff using multicasting.

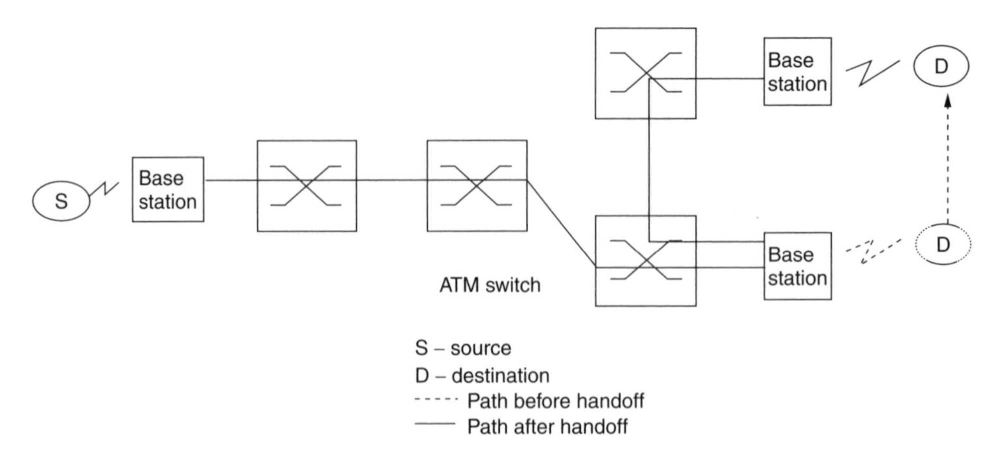

Figure 12.5 Handoff using connection extension.

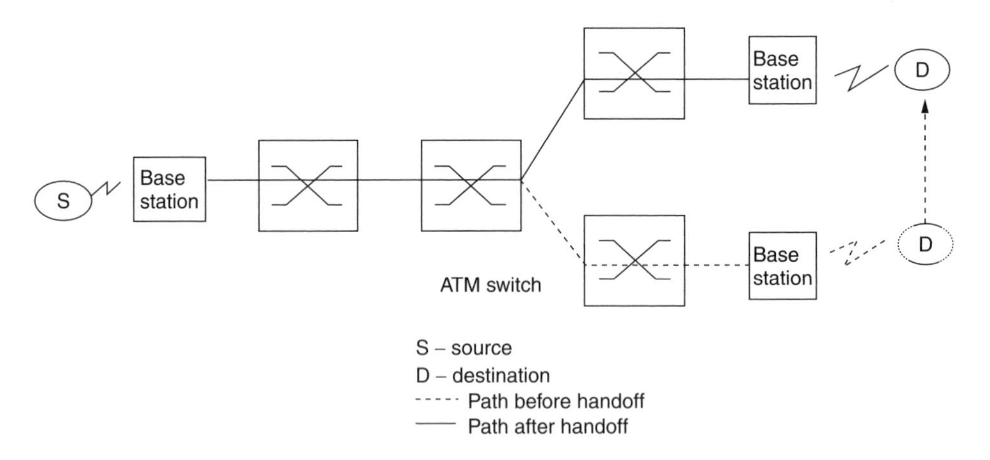

Figure 12.6 Handoff using partial reestablishment.

connection, by opening a connection to the COS. This way, it attempts to reuse as much of the existing connection as possible. The old partial path is then torn down and the resources are released. Buffering is done at the COS, which ensures in-order delivery of the cells. Figure 12.6 shows the handoff using partial reestablishment.

Handoff using two-phase protocol

A two-phase handoff protocol has been proposed by Wong and Salah, which combined the connection extension and partial reestablishment schemes.

The two-phase handoff protocol consists of two phases, that is, path extension and path optimization. Path extension is performed for each interswitch handoff. Path optimization is activated when the delay constraint or other cost is violated. Figure 12.7 shows the

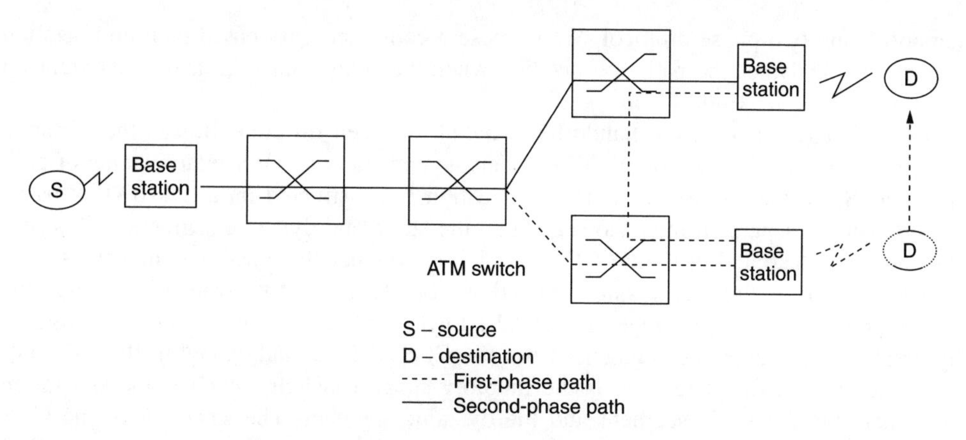

Figure 12.7 Handoff using two-phase scheme.

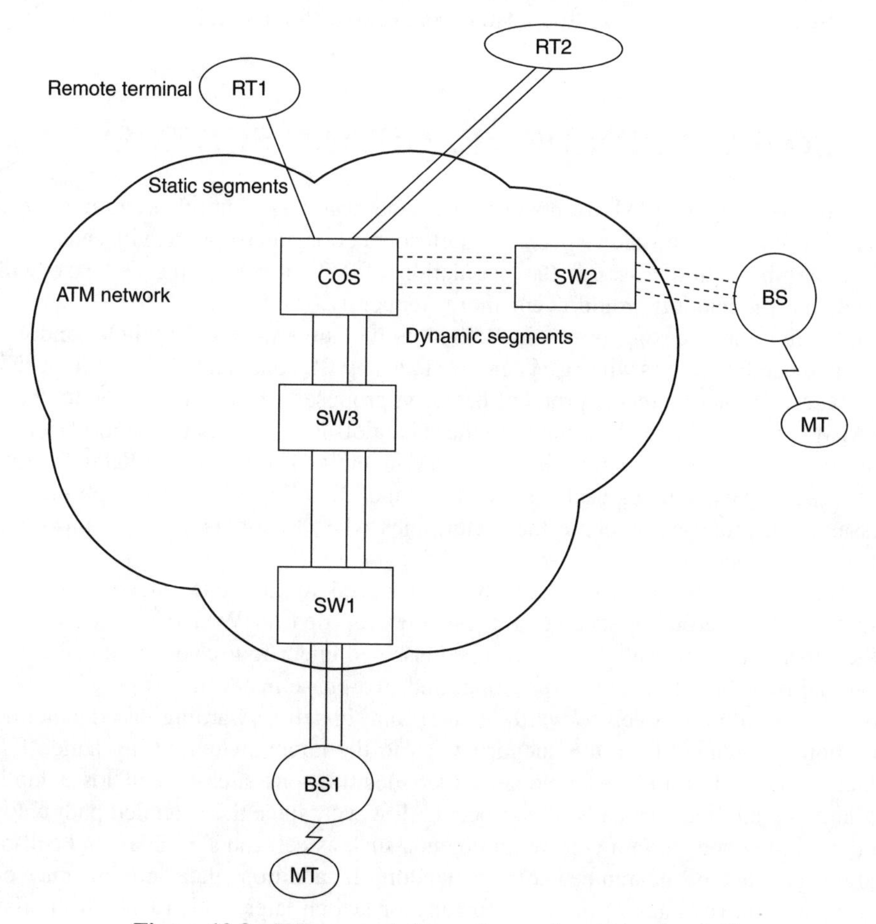

Figure 12.8 VC segmentation and rerouting during handoff.

handoff using two-phase protocol. We propose a combined QoS-based path optimization scheme that activates the path optimization when the delay constraint and path extension hops exceed the maximum value.

In a WATM network, the handoff of an MT between different BSs is the dynamic reconfiguration of the end-to-end VCs under the constraint of QoS requirements of connections. Since the end-to-end virtual connection is constructed on a link-by-link basis, the VC can be separated into two segments: the static and dynamic segments. This segmentation is illustrated in Figure 12.8 in which an MT has three active connections from two Remote Terminals (RT), one connection from RT1 and two from RT2. When the MT migrates from the coverage area of BS1 to BS2, these connections are reconfigured by creating three dynamic segments from the COS to BS2, and changing the VC routing table at the COS. The COS is the mobility support switch, which is the separation point between the static segment and the dynamic segment. The selection of the COS is mainly on the basis of the routing optimization for the connections and their QoS requirements. It is possible that there are several COSs for the connections. Here, we assume that there is one COS for all the connections for an MT in order to simplify the problem.

12.3 COMPARISON OF REROUTING SCHEMES

To support mobility in WATM networks, fast and seamless handoff is crucial. Because of the very high transmission speed, a short connection interruption will cause a large amount of information loss. As the population of MTs increase, the cell size will be reduced, and the handoff would occur more frequently in the future.

One of the major design issues in WATM is the support of interswitch handoff. An interswitch handoff occurs when a MT moves to a new BS connecting to a different switch. Recently, a two-phase handoff protocol has been proposed to support interswitch handoff in WATM networks. The aim of the two-phase handoff is to shorten the handoff delay and at the same time to use the network resources efficiently. The two-phase handoff protocol employs path extension for each interswitch handoff, followed by path optimization if necessary. We propose a scheme that determines when to trigger path optimization for the two-phase handoff.

Several connection protocols have been proposed to facilitate interswitch handoff. There are several rerouting schemes for handoff proposed for WATM networks.

The existing rerouting algorithms can be classified under four categories: cell forwarding, virtual tree–based, dynamic rerouting, and two-phase handoff.

Yuan's algorithm uses cell-forwarding algorithm. In cell-forwarding-based handoff, the connection is extended from the anchor switch to the target switch during handoff. This scheme is fast and simple to implement. QoS degradations such as cell loss, duplicate cells, and missed sequence cells do not occur. However, since the extended path is longer than the original one, certain QoS requirements, such as cell-transfer delay and cell-delay variation may not be guaranteed after a handoff. In addition, data looping may occur when the MT moves back to the previous anchor switch later, which leads to inefficient use of the network resources.

Source Routing Mobile Circuit (SRMC) algorithm uses virtual tree–based rerouting. This type of rerouting scheme creates multiple connections for a single user connection to all possible handoff candidate zones and performs an immediate rerouting. This algorithm is fast; however, it preestablishes connections to all the neighbor BSs. This may result in the waste of network resources for handoff. For example, it may waste the bandwidth, which is one of the main resources in wireless communication.

The dynamic rerouting algorithm is based on the principle of partial reestablishment of the user connection. Bora's algorithm is dynamic rerouting algorithm. In dynamic rerouting scheme, a portion of the connection is rerouted at a COS. The COS is a rerouting point in which the new partial path meets the old path. The scheme provides only partial route optimization and requires an implementation of a COS selection algorithm during handoff. The handoff latency of this scheme depends largely on selecting the COS and setting up new VCs for the establishment of the new partial path. The delay will be highly variable and will depend on the number of intermediate switches and the processing load at each switch. The delay is more noticeable in the interswitch handoff as the number of intermediate switches increase.

On the basis of the previous three schemes, a new scheme is proposed to overcome these three schemes' drawback. This new scheme is a two-phase handoff. Salah's and Wong's algorithms use this scheme. The two-phase handoff algorithm employs path extension for each interswitch handoff, followed by path optimization. The advantage of this scheme is that it provides low handoff delay and drop rate while the network resources are efficiently used.

We propose a scheme that determines when to trigger path optimization for the two-phase handoff. During the path optimization, the network determines the optimized path between the source and the destination (i.e., the path between the COS and the current anchor switch in Figure 12.9), and transfers the user information from the old path to the new path. Since the MT is still communicating over the extended path *via* the current BS

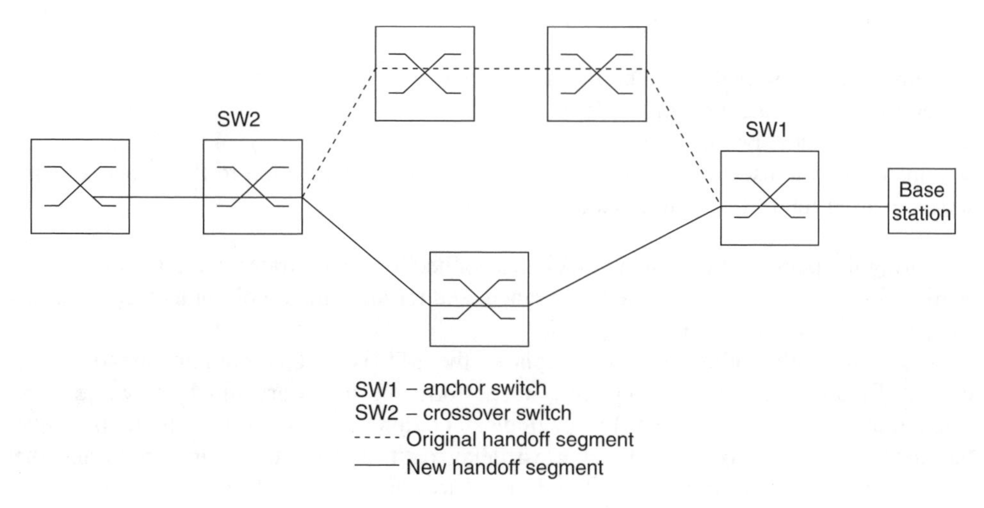

Figure 12.9 Crossover switch determination.

while path optimization takes place, this gives enough time for the network to perform the necessary functions while minimizing any service disruptions. Although path optimization can increase the network utilization by rerouting the connection to a more efficient route, transient QoS degradations such as ATM cell loss and an increase in cell-delay variation may occur. In addition, path optimization increases the processing load of certain switches and increases the signaling load of the network. Thus, path optimization after each path extension may not be necessary or desirable.

To ensure a seamless path optimization, two important issues need to be addressed:

1. How to minimize service disruptions during path optimization?
2. When and how often should path optimization be performed?

We propose a signaling protocol in which buffering is used at the COS and the anchor switch during path optimization to prevent ATM cell loss and maintain cell sequencing. We also propose a path optimization scheme, which is the QoS-based path optimization scheme. This scheme invokes the path optimization when the number of hops of the path is greater than a certain number or when the end-to-end cell-transfer delay bound is violated.

12.4 MAINTAINING THE CELL SEQUENCE DURING PATH OPTIMIZATION

The path optimization procedure consists of two phases, the path optimization initiation phase and the path optimization execution phase. The path optimization we propose is initiated on the basis of the combined QoS-based scheme. In this scheme, path optimization is initiated when the delay constraint of the link has a chance to be violated or the number of path extension hops exceed the maximum hops. The major steps during the path optimization execution phase involve

- computing a new path between the new source switch and the boundary switch;
- determining the location of COS;
- setting up a new branch connection;
- transferring the user information from the old branch connection to the new one; and
- terminating the old branch connection.

During the path rerouting, which includes connection setup, transferring the user information from the old branch to the new branch, and terminating the old branch connection, the cell sequence should be maintained.

When the path optimization takes place, the MT is still communicating over the extended path. This gives enough time to perform the necessary functions while minimizing any service disruptions. In our implementation, the optimal path is the path that has the least link delay. It is the shortest delay path between the source node and the destination node. Dijkstra's algorithm is used to compute the shortest path. The time complexity of Dijkstra's algorithm is $O(N^2)$.

After the new handoff path is computed, the source switch can determine the location of the COS. On the basis of the routing information, the source switch compares the original and the new path to determine the location of the COS. The COS is the one in which the original and the new path begin to diverge. Figure 12.9 shows an example of the determination of the COS. In Figure 12.9, the upper portion connection represents the original handoff segment, and the lower portion connection represents the new handoff segment. The original and the new handoff segments can be the same if the COS is the same as the anchor switch. In this case, the current handoff segment is the optimal path; no further path optimization executions are necessary.

When the COS is determined, connection setup between the anchor switch and the COS is performed. Connection setup is done by using the ATM UNI signaling protocol. Referring to Figure 12.10, the anchor switch first sends a signaling packet CONNECT to the COS. The COS then replies by sending CONNECT ACKNOWLEDGE (ACK) packet to the anchor switch. This message indicates that the connection setup of the new handoff segment between the COS and the anchor switch is successful. Now there are two segments established between the COS and the anchor switch. The connection rerouting protocol has to reroute the data flow from the original handoff segment to the new handoff segment. We introduce the procedure of transfer user information in the uplink and downlink to the new handoff segment in the following subsections.

The path optimization signaling protocol ensures that ATM cell sequencing is maintained on the uplink and downlink by buffering cells at the anchor switch and the COS, respectively. The procedure of transferring user information in the uplink is as follows:

1. In the uplink, the anchor switch sends the COS a LAST(m) message, which indicates that no more ATM cells from the MT m will be transmitted through the original uplink handoff segment.
2. The anchor switch then begins to transmit ATM cells through the new handoff segment.
3. The COS will not transmit any ATM cells from the new uplink segment until it has received the LAST(m) message from the original uplink handoff segment. This ensures the uplink ATM cells sequencing, since the delay of the new handoff segment is shorter than the delay of the original handoff segment. If the COS transmits ATM cells from the new handoff segment without receiving the LAST(m) message, the later ATM cells from the new handoff segment, which arrives first, will transmit first. The cell sequencing is then disrupted. Buffers can be used at the COS to temporarily store the cells arriving from the new uplink handoff segments until the LAST(m) message is received.
4. When the COS receives the LAST(m) packet, it begins to transmit ATM cells from the new uplink handoff segment.
5. The COS sends a LAST ACK(m) packet to the anchor switch to acknowledge the reception of the LAST(m) packet.

The procedure of transfer user information in the downlink is as follows:

1. In the downlink, the COS sends the anchor switch a LAST(m) message, which indicates that no more ATM cells that correspond to MT m will be transmitted through the original downlink handoff segment.

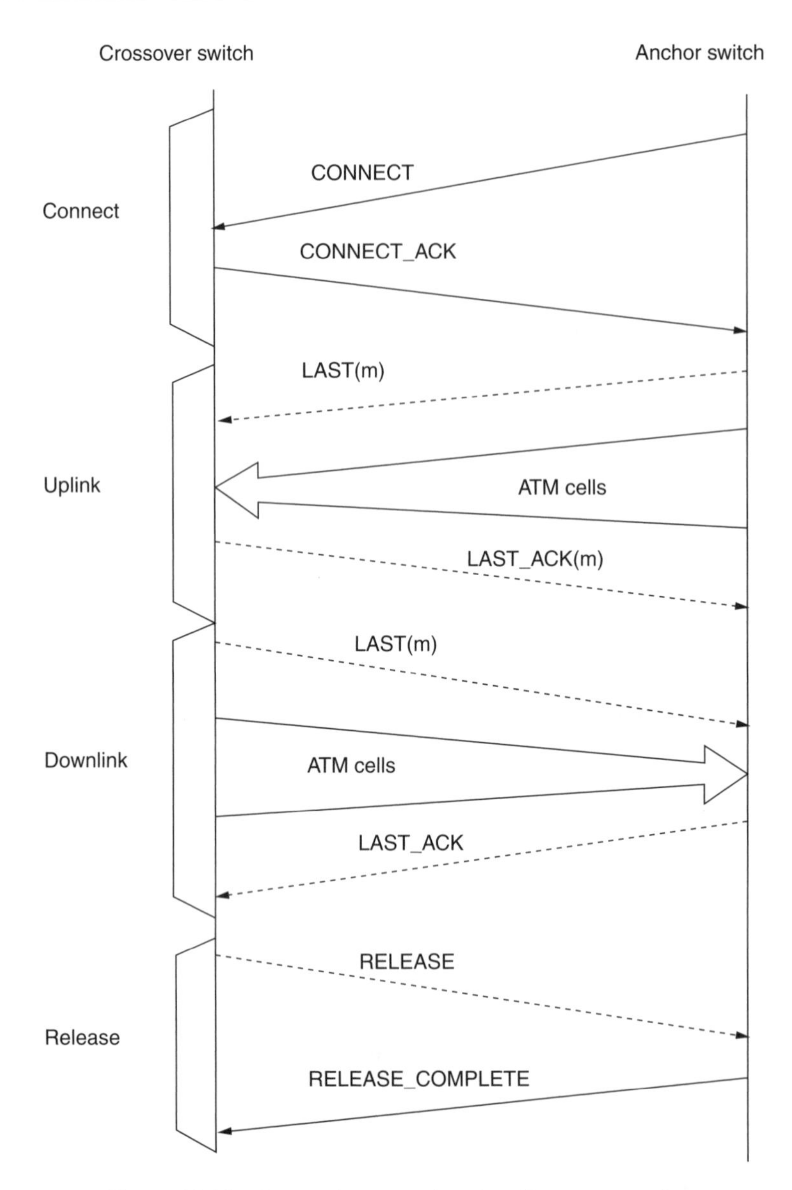

Figure 12.10 Connection rerouting signaling message flow.

2. The COS then begins to transmit ATM cells through the new handoff segment.
3. The anchor switch will not transmit any ATM cells from the new uplink segment until it has received the LAST(m) message from the original uplink handoff segment. This ensures the downlink ATM cell sequencing, since the delay of the new handoff segment is shorter than the delay of the original handoff segment. If the anchor switch transmits ATM cells from the new handoff segment without receiving the LAST(m)

message, the later ATM cells from the new handoff segment, which arrive first, will be transmitted first. The cell sequencing is then disrupted. Buffers can be used at the anchor switch to temporarily store the cells arriving from the new downlink handoff segments until the LAST(m) message is received.
4. When the anchor switch receives the LAST(m) packet, it begins to transmit ATM cells from the new downlink handoff segment.
5. The anchor switch sends a LAST ACK(m) packet to the anchor switch to acknowledge the reception of the LAST(m) packet.

Once the connection rerouting process is completed, the COS sends a signaling packet RELEASE to terminate the original segment between the COS and the anchor switch. The anchor switch sends a RELEASE COMPLETE packet to the COS after it receives the RELEASE packet. The RELEASE COMPLETE packet indicates that the original handoff segment has been released. The path optimization process is now complete.

By employing the procedure discussed above, the cell sequence during the rerouting of a user connection due to a handoff attempt is preserved. Occasionally, cell sequence will be broken in the WATM network. Recovery from such sequencing errors is the responsibility of the upper-layer protocols, such as the AAL, and sometimes the user application layer. For data traffic, recovery may be attempted by retransmission. For time-sensitive traffic, the sequence error may need to be corrected by voice or video interpolation techniques.

12.5 COMBINED QoS-BASED PATH OPTIMIZATION SCHEME

The end-to-end path after path extension may not be optimum. These paths may lead to an inefficient use of network resources. There are different methods to initiate path optimization. Path optimization schemes can be classified into four types, that is, QoS-based, network-based, time-based, and handoff-based.

QoS-based path optimization schemes trigger path optimization of each mobile connection on the basis of its current QoS measures. For example, path optimization can be initiated if the number of hops of the path is greater than a certain number, or if the end-to-end cell-transfer delay bound is violated. To implement those QoS-based path optimization schemes, information about the quality of the current path in terms of the defined QoS measurement (e.g., hop count, current average delay) must be maintained by the network. Wong proposed a QoS-based path optimization in which the path optimization is initiated when the delay constraint of the handoff segment is violated. This scheme ensures that the delay of the mobile connection is always below the specified delay constraint. Chan proposed a scheme in which the path optimization is initiated when the number of hops by which the path is extended is greater than the maximum number of hops. This improves the path efficiency and at the same time allows path rerouting to efficiently utilize the network resources.

In the time-based path optimization scheme, the path optimizations are triggered at time instants that are independent of the current QoS of the connection or the utilization of the network. The time instants can be deterministic or random. For example, the time between path optimization can be based on some random processes. In addition, it can also be a function of the velocity of the MT, the dwell time, and the residual service time of the mobile connection.

In the handoff-based path optimization scheme, the path optimizations are triggered on the basis of some criteria after each interswitch handoff. Thus, it can be a function of the number of previous handoff, the velocity of the MT, and the residual service time. Wong proposed and analyzed three path optimization schemes: exponential, periodic, and Bernoulli. These schemes provide the means of optimizing the frequency of path optimization with respect to the average cost per call. An analytical model and a simulation model are used to analyze the performance of the path optimization schemes. The simulation result shows that Bernoulli path optimization scheme out performs the other two schemes by giving the lowest expected cost per call.

Path optimization can be invoked on the basis of a combination of the above schemes. We propose a QoS-based scheme that combines Wong's and Chan's schemes. We compared the performance of these schemes and our combined QoS-based path optimization scheme by using the following measures: handoff drop rate, average number of hops, average link delay, and rerouting delay for handoff.

This scheme invokes the path optimization when the number of path extensions is greater than a certain number or the end-to-end cell-transfer delay bound is violated. This scheme ensures that during the handoff each hop meets the delay variation so that the drop rate variable is improved. Also, even if the delay variation is not violated, the path optimization scheme is invoked after a certain number of path extensions. This ensures the efficient utilization of the network resources and avoids the loop.

We propose a combined QoS-based path optimization algorithm to improve the handoff drop rate. In this section, we compare the QoS-based path optimization algorithm with the existing delay-based path optimization algorithm and hop-based path optimization algorithm. We describe the algorithm of these three path optimization algorithms as follows:

In the combined QoS-based path optimization algorithm, when there is a handoff request, the path extension is done first to ensure a seamless handoff. After the path extension, the new path's delay is the sum of the previous path delay and the delay between the new source node and the previous source node. Here we assume the destination node is stationary during handoff. We also assume the previous path does not exceed the maximum delay limit. After the path extension rerouting, the delay of the path may exceed the required maximum delay value. If the delay variation is violated, the path optimization is activated. The new shortest path, that is, the lowest delay path should be found to substitute the previous path to ensure a continuous connection. Also, after the path extension rerouting, the total hops of the path may not be optimum, or may have loops. This situation is caused by the large number of path extensions. The combined QoS scheme also checks if the number of path extensions exceeds the certain given number. If it does, then the path optimization is also activated. The given number of path extensions is based on the system design. We use four hops in the simulation. That means path optimization is activated when the number of path extensions exceeds the four hops. This number is based on Chan's proposal.

The idea of the delay-based path optimization algorithm is that the path extension continues until the delay variation is violated during handoff. When a MT first requests handoff during the call, the previous path is reserved. The rerouting path is the extension of the previous path. That means the rerouting path is extended from the previous switch to the current switch. This is the simplest and the fastest way to do path rerouting and it ensures a seamless handoff. After the path extension is completed, the new path's delay is the sum of the previous path delay and the delay between the new source node and the previous source node. Here we assume the destination node is stationary during handoff. We also assume the previous path does not exceed the maximum delay limitation. After path extension rerouting, the delay of the path may exceed the required maximum delay value. This may cause the call to drop.

If the delay variation is violated, the path optimization is activated. The new shortest path, that is, the lowest delay path should be found to substitute the previous path to ensure a continuous connection.

In the hop-based algorithm, the path extension is done when there is a handoff request until the number of path extensions exceeds the maximum allowed path extension hops. When the number of path extensions exceeds the maximum allowed path extension hops, path optimization is activated when there is a handoff request. This algorithm only considers the number of path extensions. It does not consider the delay variation to activate the path optimization scheme.

The rerouting path searching procedure for delay-based, hop-based, and combined rerouting algorithms is the shortest path search.

The delay-based algorithm extends the path each time when handoff occurs until the link delay is larger than the allowed maximum delay. In our simulation, we use MAXDELAY to indicate the allowed maximum delay. When the MAXDELAY is large, which means the traffic is not time critical, it ensures the fastest handoff and lower handoff drop rate. But the number of hops of the path is not optimum. Since it extends the path each time when handoff is activated, it may not be the shortest path anymore. Also, the loop may occur when the MT moves from the new source node to the original source node again after the handoff. This causes a waste of network resources.

The hop-based algorithm activates path rerouting when it has maximum allowed hops of path extensions no matter whether the delay variation is violated. This algorithm works well when the maximum allowed delay (i.e., MAXDELAY) is set large. But when the MAXDELAY is small, the probability of the handoff drop is high. Since each path extension adds the weight between new source node and old source node to the previous total weight of the link, it means the link delay is increased each time. If the link delay is larger than the MAXDELAY, the call is dropped. Thus, when the MAXDELAY is small, the handoff drop rate will be higher than the delay-based algorithm.

The combined QoS-based algorithm combines the delay-based and hop-based algorithms into one algorithm. The combined algorithm checks the delay variation every time the MT has a handoff request. This takes advantage of the delay-based rerouting algorithm, which ensures faster handoff and lower handoff drop rate than the hop-based rerouting algorithm. The difference between the combined algorithm and the delay-based algorithm is that the combined algorithm will activate the path rerouting when the path extension reaches four hops. This takes advantage of the hop-based handoff algorithm, which ensures the optimized path and effectively saves network resources. Also, the average hops for the handoff request is lower than the delay-based algorithm.

The disadvantage of the combined algorithm is that it has an overhead since it needs to check the delay variation for every handoff request. So it takes longer to handle the handoff. The trade-off is the high reliability, low drop rate, and high utilization of network resources.

These three rerouting algorithms are all sensitive to the network topology. The complexity of each of these three algorithms is $O(N^2)$, where N is the number of nodes in the network.

12.6 SUMMARY

The QoS-based rerouting algorithm is designed to implement two-phase interswitch handoff scheme for WATM networks. A path extension is used for each interswitch handoff, and the path optimization is invoked when the handoff path exceeds the delay constraint or the maximum path extension hops constraint. There are different types of path optimization schemes: combined QoS-based, delay-based, hop-based path rerouting schemes, and QoS combined path optimization scheme for WATM network. The QoS combined path optimization scheme focuses on the problems related to the support of mobility in the WATM network. This scheme determines when to trigger path optimization for the two-phase handoff and when to minimize service disruption during path optimization.

PROBLEMS TO CHAPTER 12

Two-phase combined QoS-based handoff scheme

Learning objectives

After completing this chapter, you are able to

- demonstrate an understanding of a WATM network;
- explain hard and soft handoff;
- explain forward and backward handoff;
- explain combined QoS-based path optimization scheme; and
- explain different types of path optimization schemes.

Practice problems

12.1: What are the major components in a WATM network?
12.2: How is a hard handoff executed?
12.3: How is a soft handoff executed?
12.4: How is a forward handoff performed?
12.5: How is a backward handoff performed?
12.6: What is a handoff using full reestablishment?

12.7: What is a handoff using multicasting?
12.8: What is a handoff using connection extension?
12.9: What is a handoff using partial reestablishment?
12.10: What is a handoff using two-phase protocol?
12.11: What is a combined QoS-based path optimization scheme?
12.12: What are the types of path optimization schemes?

Practice problem solutions

12.1: There are two major components in a WATM network: a radio access layer providing high-bandwidth wireless transmission with appropriate MAC, DLC, and so on and a mobile ATM network for interconnection of BSs (APs) with appropriate support of mobility related functions, such as handoff and location management.

12.2: In a hard handoff, the MT switches the communication from the old link to the new link. Thus, there is only one active connection from the MT at any time. There is a short interruption in the transmission. This interruption should be minimized in order to make the handoff seamless.

12.3: In a soft handoff, the MT is connected simultaneously to two APs. As it moves from one cell to another, it 'softly' switches from one BS to another. When connected to two BSs, the network combines information received from two different routes to obtain better quality. This is commonly referred to as macrodiversity.

12.4: In a forward handoff, after the MT decides the cell to which it will make a handoff, it contacts the BS controlling the cell. The new BS initiates the handoff signaling to link the MT from the old BS. This is especially useful if the MT suddenly loses contact with the current BS.

12.5: In a backward handoff, after the MT decides the cell to which it attempts to make a handoff, it contacts the current BS, which initiates the signaling to handoff to the new BS.

12.6: A handoff using full reestablishment occurs in a connection-oriented wireless environment, in which virtual circuits are established from the source to the destination. The data follows the path that has been set up, and an in-order delivery is guaranteed. If a handoff is to occur, the old virtual connection is torn down, and an entirely new virtual circuit is set up from the current source to the current destination. Since both ends are explicitly involved, this handoff scheme is not transparent. Severe traffic interruptions are experienced and hence this scheme is not recommended.

12.7: A handoff using multicasting is used in both the connection-oriented and connectionless scenarios. In the case of a WATM environment, multicasting is used to establish links to all BSs that are neighboring the BS that is currently controlling a MT. Subsequently, in whichever direction the MT moves, a handoff path has already been established. Also, since the data is being multicast, it continues to flow without any interruption. This scheme ensures a lossless and seamless handoff. However, since data is being multicast to the entire set of nodes most of which is unused, bandwidth is being utilized very inefficiently. Also, if an MT is at the edge of two cells, it is very likely that it might get two copies of the data packets. This leads to other complications like BS synchronization.

12.8: The basic idea of the handoff using connection extension scheme is that the local paths are more affordable than the global paths. When an MT migrates from one BS to another, the old BS extends the connection to the new BS. The obvious disadvantage of this method is that the new path to the MT is not an optimal path.

12.9: A handoff using partial reestablishment uses the concept of a COS. The new BS does a partial reestablishment of the connection by opening a connection to the COS. This way it attempts to reuse as much of the existing connection as possible. The old partial path is then torn down and the resources are released. Buffering is done at the COS.

12.10: A handoff using two-phase protocol combines the connection extension and partial reestablishment schemes. The two-phase handoff protocol consists of two phases: path extension and path optimization. Path extension is performed for each inter-switch handoff. Path optimization is activated when the delay constraint or other cost is violated.

12.11: A combined QoS-based path optimization scheme activates the path optimization when the delay constraint and path extension hops exceed a maximum value.

In the combined QoS-based path optimization algorithm, when there is a handoff request, the path extension is done first to ensure a seamless handoff. After the path extension, the new path's delay is the sum of the previous path delay and the delay between the new source node and previous source node.

12.12: Path optimization schemes can be classified into four types: QoS-based, network-based, time-based, and handoff-based. QoS-based path optimization schemes trigger path optimization of each mobile connection on the basis of its current QoS measures.

References

A. S. Acampora and M. Naghshineh (1994) An architecture and methodology for mobile-executed handoff in cellular ATM networks. *IEEE JSAC*, **12**, 1365–1375.

I. F. Akyildiz, J. McNair, J. Ho, H. Uzunalioglu and W. Wang (1998) Mobility management in current and future communications networks. *IEEE Network*, **12**(4), 39–49.

I. Akyildiz, H. Uzunalioglu and M. D. Bender (1999) Handover management in low earth orbit (LEO) satellite networks. *Mobile Networks and Applications*, **4**, 301–310.

B. A. Akyol and D. C. Cox Rerouting for Handoff in a Wireless ATM Network International Conference on Universal Personal Communications, 1996, pp. 374–379. URL: http://wireless.stanford.edu/~akyol.

G. Anastasi, L. Lenzini, E. Mingozzi, A. Hettich and A. Kramling (1998) MAC protocols for wideband wireless local access: evolution toward wireless ATM. *IEEE Personal Communications*, **5**(5), 53–64.

A. A. Androutsos, T. K. Apostolopoulos and V. C. Daskalou (2000) Managing the network state evolution over time using CORBA environment. *IEEE Journal on Selected Areas in Communications*, **18**(5), 654–663.

E. Ayanoglu (1999) Wireless broadband and ATM systems. *Computer Networks*, **31**, 395–409.

S. Batistatos, K. Zygourakis, F. Panken, K. Raatikainen and S. Trigila (1999) TINA architecture extensions to support terminal mobility. *Proc. IEEE Telecommunications Information Networking Architecture Conference*, pp. 34–45.

S. Boumerdassi and A.-L. Beylot (1999) Adaptive channel allocation for wireless PCN. *Mobile Networks and Applications*, **4**, 111–116.

J. Bray and C. F. Sturman (2000) *Bluetooth: Connect Without Cables*, Prentice Hall PTR, Upper Saddle River, NJ.

M. Breugst, L. Faglia, O. Pyrovolakis, I. S. Venieris and F. Zizza (1999) Towards mobile service agents within and advanced broadband IN environment. *Computer Networks*, **31**(19), 2037–2052.

J. Broch, D. Johnson and D. Maltz (1998) The Dynamic Source Routing Protocol for Mobile Ad Hoc Networks, Internet Draft, draft-ietf-manet-dsr-00.txt, March 13 1998; Work in progress.

J. Broch, D. Maltz, D. B. Johnson, Y.-C. Hu and J. Jetcheva (1998) A performance comparison of multi-hop wireless ad hoc network routing protocols. *Proceedings of the Fourth Annual ACM/IEEE International Conference on Mobile Computing and Networking (MobiCom '98)*, October 25, 1998.

A. T. Campbell, R. R.-F. Liao and Y. Shobatake Supporting QoS Controlled Handoff in Mobiware. URL: http://www.ctr.columbia.edu/~liao/project/hdoff/signal_abstract.html.

K. S. Chan, S. Chan and K. T. Ko (1998) Hop-limited handoff scheme for ATM-based broadband cellular networks. *Electronics Letters*, **34**(1), 26–27.

K.-S. Chan and S. Chan (2000) An efficient handoff management scheme for mobile wireless ATM networks. *IEEE Transactions on Vehicular Technology*, **49**(3), 799–815.

F. G. Chatzipapadopoulos, M. K. Perdikeas and I. S. Venieris (2000) Mobile agent and CORBA technologies in the broadband intelligent network. *IEEE Communications Magazine*, **38**(6), 116–124.

T.-W. Chen and M. Gerla Global state routing: a new routing scheme for ad-hoc wireless networks. *Proc. IEEE ICC '98*, 1998, pp. 171–175.

I.-R. Chen, T.-M. Chen and C. Lee (1998) Agent-based forwarding strategies for reducing location management cost in mobile networks. *Proc. IEEE International Conference on Parallel and Distributed Systems*, pp. 266–273.

C.-C. Chiang (1997) Routing in clustered multihop. *Mobile Wireless Networks with Fading Channel Proc. IEEE SICON '97*, April 1997, pp. 197–211.

J. Chuang and N. Sollenberger (1999) Wideband wireless data access based on OFDM and dynamic packet assignment. *IEEE Wireless Communications and Networking Conference*, **2**, 757–761.

Composite Capabilities/Preference Profiles: Requirements and Architecture, W3C Working Draft, July 21 2000. URL: http://www.w3.org.

Concordia. URL: http://www.meitca.com/HSL/Projects/Concordia.

S. Corson, S. Papademetriou, P. Papadopoulos, V. Park and A. Qayyum (1998) An Internet MANET Encapsulation Protocol (IMEP) Specification, Internet Draft, draft-ietf-manet-imep-spec-01.txt, August 7 1998; Work in progress.

G. Coulson (1999) A configurable multimedia middleware platform. *IEEE Multimedia*, **6**(1), 62–76.

A. Dornan (2000) *The Essential Guide to Wireless Communications Applications*, Prentice Hall PTR, Upper Saddle River, NJ.

C. Dou, Y.-P. Chen and H.-K. Chen (2001) An agent-based platform for dynamic service provisioning in 3G mobile systems: scenarios and performance analyses. *Proc. 15th IEEE International Conference on Information Networking*, pp. 883–888.

R. Dube, C. Rais, K.-Y. Wang and S. Tripathi (1997) Signal stability based adaptive routing (SSA) for ad-hoc mobile networks. *IEEE Personal Communications*, **4**(1), 36–45.

H. J. R. Dutton and P. Lenhard (1995) *Asynchronous Transfer Mode*, Prentice Hall, New York.

W. K. Edwards (1999) *Core Jini*, Prentice Hall PTR, Upper Saddle River, NJ.

K. Y. Eng, M. J. Karol, M. Veeraraghavan, E. Ayanoglu, C. B. Woodworth, P. Pancha and R. A. Valenzuela (1995) BAHAMA: A broadband ad-hoc wireless ATM local area network. *ICC '95*, pp. 1216–1223.

Extensible Markup Language (XML) 1.0, W3C Recommendation, October 6 2000. URL: http://www.w3.org.

A. Fladenmuller and R. De Silva (1999) The effect of mobile IP handoffs on the performance of TCP. *Mobile Networks and Applications*, **4**, 131–135.

V. K. Garg (2000) *IS-95 CDMA and cdma 2000*, Prentice Hall PTR, Upper Saddle River, NJ.

R. Ghai and S. Singh (1994) An architecture and communication protocol for picocellular networks. *IEEE Personal Communication*, **1**(3), 36–46.

Z. Haas and M. Pearlman (1998) The Zone Routing Protocol (ZRP) for Ad Hoc Networks, Internet Draft, draft-ietf-manet-zone-zrp-01.txt, August, 1998; Work in progress.

A. Hać (2000) *Multimedia Applications Support for Wireless ATM Networks*, Prentice Hall PTR, Upper Saddle River, NJ.

A. Hać and B. L. Chew (2000) Demand assignment multiple access protocols for wireless ATM networks. *Proc. IEEE Vehicular Technology Conference*, Boston, MA, September 24–28 2000, pp. 237–241.

A. Hać and B. L. Chew (2001) ARCMA–adaptive request channel multiple access protocol for wireless ATM networks. *International Journal of Network Management*, **11**(6), 333–363.

A. Hać and J. Peng (2001) A two-phase combined QoS-based handoff scheme in wireless ATM network. *International Journal of Network Management*, **11**(6), 309–330.

A. Hać and J. Peng (2001) Handoff in wireless ATM network. *International IEEE Conference on Third Generation Wireless and Beyond*, San Francisco, CA, May 30–June 2, 2001, pp. 894–899.

A. Hać and Y. Zhang (2002) Reducing signaling traffic in wireless ATM networks through handoff scheme improvement. *International Journal of Network Management*, **12**(5), in press.

A. Hać and Y. Zhang (2001) Signaling QoS in wireless ATM network. *International IEEE Conference on Third Generation Wireless and Beyond*, San Francisco, CA May 30–June 2, 2001, pp. 906–911.

A. Hać and Z. Zhu (1999) Performance of routing schemes in wireless personal networks. *International Journal of Network Management*, **9**(2), 80–105.

L. Hagen, J. Mauersberger and C. Weckerle (1999) Mobile agent based service subscription and customization using the UMTS virtual home environment. *Computer Networks*, **31**(19), 2063–2078.

M. Hannikainen, J. Knuutila, A. Letonsaari, T. Hamalainen, J. Jokela, J. Ala-Laurila and J. Saarinen (1998) TUTMAC: A medium access control protocol for a new multimedia wireless local area network. *Proc. Ninth IEEE International Symposium on Personal, Indoor and Mobile Radio Communications*, **2**, 592–596.

C. Hedrick (1988) Routing Information Protocol, RFC 1058, June 1988.

IBM: Aglets Software Development Kit. URL: http://www.trl.ibm.co.jp/aglets.

M. Inoue, G. Wu and Y. Hase (1999) IP-based high-speed multimedia wireless LAN prototype for broadband radio access integrated network (BRAIN). *Proc. 50th IEEE Vehicular Technology Conference*, **1**, 357–361.

International Standard, ISO/IEC 8802-11, ANSI/IEEE, Std 802.11, 1999.

Internet Protocol, RFC 791, Information Sciences Institute, September 1981.

A. Iwata, C.-C. Ching, G. Pei, M. Gerla and T.-W. Chen (1999) Scalable routing strategies for ad hoc wireless networks. *IEEE Journal on Selected Areas in Communications*, **17**(8), 1369–1379.

P. Jacquet, P. Muhlethaler and A. Qayyum (1998) Optimized Link State Routing Protocol, Internet Draft, draft-ietf-manet-olsr-00.txt, November 18 1998; Work in progress.

B. Jennings, R. Brennan, R. Gustavsson, R. Feldt, J. Pitt, K. Prouskas and J. Quantz (1999) FIPA-compliant agents for real-time control of intelligent network traffic. *Computer Networks*, **31**(19), 2017–2036.

M. Jiang, J. Li and Y. C. Tay (1998) Cluster Based Routing Protocol (CBRP) Functional Specification, Internet Draft, draft-ietf-manet-cbrp-spec-00.txt, August 1998; Work in progress.

M. Jiang, J. Li and Y. C. Tay (1999) Cluster Based Routing Protocol, August 1999, IETF Draft. URL: http://www.ietf.org/internet-drafts/draft-ietf-manet-cbrp-spec-01.txt.

M. Joa-Ng and I.-T. Lu (1999) A peer-to-peer zone-based two-level link state routing for mobile ad hoc networks. *IEEE Journal on Selected Areas in Communications*, **17**(8), 1415–1425.

D. Johnson and D. Maltz (1996) Truly seamless wireless and mobile host networking. Protocols for adaptive wireless and mobile networking. *IEEE Personal Communications*, **3**(1), 34–42.

D. B. Johnson (1994) Routing in ad hoc networks of mobile hosts. *Proceedings of the IEEE Workshop on Mobile Computing Systems and Applications*, December, 1994.

D. Johnson (1996) Dynamic source routing in ad hoc wireless networks, *Mobile Computing*, Kluwer Academic Publishers, Boston, MA.

D. B. Johnson and D. A. Maltz (1996) Dynamic source routing in ad hoc networks. In *Mobile Computing*, T. Imielinski and H. Korth (eds), Kluwer Academic Publishers, Boston, MA, pp. 152–181.

D. B. Johnson and D. A. Maltz (1999) The Dynamic Source Routing Protocol for Mobile Ad Hoc Networks, October 1999, IETF Draft. URL: http://www.ietf.org/internet-drafts/draft-ietf-manet-dsr-03.txt.

A. Kalaxylos, S. Hadjiefthymiades and L. Merakos (1998) Mobility management and control protocol for wireless ATM networks. *IEEE Network*, **12**(4), 19–27.

M. R. Karim (2000) *ATM Technology and Services Delivery*, Prentice Hall PTR, Upper Saddle River, NJ, pp. 141–256.

M. Karol, Z. Liu and K. Y. Eng (1995) Distributed-queuing request update multiple access (DQRUMA) for wireless packet (ATM) networks. *Proc. IEEE ICC International Conference on Communications*, Seattle, WA, June 1995, pp. 1224–1231.

J. Keogh (2000) *The Essential Guide to Networking*, Prentice Hall PTR, Upper Saddle River, NJ.

P. Kumar and L. Tassiulas (2000) Mobile multi-user ATM platforms: architectures, design issues, and challenges. *IEEE Network*, **14**(2), 42–50.

T. Kurz, J.-Y. Le Boudec and H. J. Einsiedler (1998) Realizing the benefits of virtual LANs by using IPv6. *Proc. 1998 IEEE International Zurich Seminar on Broadband Communications*, pp. 279–283.

J. Landru, H. Mordka and P. Vincent (1998) MONACO – modular open network agent for control operations, *Proc. IEEE Network Operations and Management Symposium*, **2**, 600–609.

D. C. Lee, D. L. Lough, S. F. Midkiff, N. J. Davis, IV and P.E. Benchoff (1998) The next generation of the Internet: aspects of the Internet protocol version 6. *IEEE Network*, **12**(1), 28–33.

M. Lerner, G. Vanecek, N. Vidovic and D. Vrsalovic (2000) *Middleware Networks*, Kluwer Academic Publishers, Boston, MA.

B. C. Lesiuk Ad Hoc Networks. URL: http://ghost.hn.org.

K. H. Le, S. Norskov, L. Dittmann and U. Gliese (1998) Base station MAC with APRMA protocol for broadband multimedia ATM in micro/pico-cellular mobile Networks. *Proc. 48th IEEE Vehicular Technology Conference*, **1**(1), 234–238.

S. Lipperts and A. S.-B. Park (1999) An agent-based middleware – a solution for terminal and user mobility. *Computer Networks*, **31**(19), 2053–2062.

Microsoft Technologies: Distributed Component Object Model. URL: http://www.microsoft.com/com/tech/dcom.asp.

B. A. Miller and C. Bisdikian (2000) *Bluetooth Revealed*, Prentice Hall PTR, Upper Saddle River, NJ.

S. Murthy and J. J. Garcia-Luna-Aceves, An Efficient Routing Protocol for Wireless Networks, ACM Mobile Networks and Application Journal, Special Issue on Routing in Mobile Communication Networks, October 1996, pp. 183–97.

H. Ohtsuka, T. Oono, Y. Kondo, O. Nakamura and T. Tanaka (1999) Potential wireless technologies in mobile communications. *Proc. IEEE International Conference on Communications*, **2**, 1121–1125.

K. Pahlavan, P. Krishnamurthy, A. Hatami, M. Ylianttila, J. P. Makela, R. Pichna and J. Vallstron (2000) Handoff in hybrid mobile data networks. *IEEE Personal Communications*, **7**(2), 34–47.

V. D. Park and M. S. Corson (1997) A highly adaptive distributed routing algorithm for mobile wireless networks. *Proc. INFOCOM '97*, April 1997.

V. Park and S. Corson (1998) A performance comparison of TORA and ideal link state routing. *Proceedings of IEEE Symposium on Computers and Communication '98*, June, 1998.

V. Park and S. Corson (1998) Temporally-ordered routing algorithm (TORA) version 1 functional specification, Internet Draft, draft-ietf-manet-tora-spec-01.txt, August 7 1998; Work in progress.

M. K. Perdikeas, F. G. Chatzipapadopoulos, I. S. Venieris and G. Marino (1999) Mobile agent standards and available platforms. *Computer Networks*, **31**(19), 1999–2016.

C. Perkins and E. Royer (1998) Ad Hoc On Demand Distance Vector (AODV) Routing, Internet Draft, draft-ietf-manet-aodv-02.txt, November 20 1998; Work in progress.

C. E. Perkins and P. Bhagwat (1994) Highly dynamic destination-sequenced distance-vector routing (DSDV) for mobile computers. *Computer Communication Review*, **24**(4), 234–244.

C. E. Perkins and P. Bhagwat (1994) Highly dynamic destination-sequenced distance-vector routing (DSDV) for mobile computers. *Proceedings of the SIGCOMM 94 Conference on Communications, Architectures, Protocols and Applications*, August, 1994, pp. 234–244.

C. E. Perkins, E. M. Royer and S. R. Das (1999) Ad Hoc On-demand Distance Vector Routing, October 1999, IETF Draft. URL: http://www.ietf.org/internet-drafts/draft-ietf-manet-aodv-04.txt.

D. Petras and A. Kramling (1997) Wireless ATM: performance evaluation of DSA++ MAC protocol with fast collision resolution by a probing algorithm. *International Journal of Wireless Information Networks*, **4**(4), 215–223.

D. Prevedourou, K. Zygourakis, S. Efremidis, G. Stamoulis, D. Kalopsikakis, A. Kyrikoglou, V. siris, M. Anagnostou, L. Tzifa, T. Louta, P. Demestichas, N. Liossis, A. Kind, K. Valtari, H. Jormakka and T. Jussila (1999) Use of agent technology in service and retailer selection in a personal mobility context. *Computer Networks*, **31**(19), 2079–2098.

R. Ramjee, T. F. La-Porta, J. Kurose and D. Towsley (1998) Performance evaluation of connection rerouting schemes for ATM-Based wireless networks. *IEEE/ACM Transactions on Networking*, **6**(3), 249–261.

Resource Description Framework (RDF) 1.0, W3C Candidate Recommendation, March 27 2000. URL: http://www.w3.org.

E. M. Royer and C.-K. Toh (1999) A review of current routing protocols for ad hoc mobile wireless networks. *IEEE Personal Communications*, **6**(2), 46–55.

I. Rubin and C. W. Choi (1997) Impact of the location area structure on the performance of signaling channels in wireless cellular networks. *IEEE Communication Magazine*, **35**(2), 108–115.

M. Sadiku and M. Ilyas (1995) *Simulation of Local Area Networks*, CRC Press, FL, Boca Raton.

K. Salah and E. Drakoponlos (1998) A Two-phase Inter-Switch Handoff scheme for wireless ATM networks. *IEEE ATM '98 Workshop Proceedings*, pp. 708–713.

D. C. Schmidt and F. Kuhns (2000) An overview of the real-time CORBA specification. *IEEE Computer*, **33**(6), 56–63.

J. Sharony (1996) An architecture for mobile radio networks with dynamically changing topology using virtual subnets. *Journal of Special Topics in Mobile Networks and Applications (MONET)* **1**(1), pp. 75–86.

J. Sharony (1996) A mobile radio network architecture with dynamically changing topology using virtual subnets. *Proceedings of ICC '96*, pp. 807–812.

B. S. Suh and Shin (1998) Dynamic scope location tracking algorithm for mobile PNNI scheme in wireless ATM networks. *Electronics Letters*, **34**(17), 1643–1644.

T. Sukuvaara, P. Mahonen and T. Saarinen (1999) Wireless internet and multimedia services support through two-layer LMDS system. *IEEE International Workshop on Mobile Multimedia Communications*, pp. 202–207.

A. S. Tanenbaum (1996) *Computer Networks*, Prentice Hall, New York, pp. 348–352.

C.-K. Toh (1996) A novel distributed routing protocol to support ad hoc mobile computing. *Proc. 1996 IEEE 15th Annual Int'l. Phoenix Conf. Comp. and Commun.*, March 1996, pp. 480–486.

C.-K. Toh (1997) *Wireless ATM and AD-HOC Networks*, Kluwer Academic Publishers, Boston, MA.

A. Umar (1997) *Object-Oriented Client/Server Internet Environments*, Prentice Hall PTR, Upper Saddle River, NJ.

M. Van der Heijden and M. Taylor (2000) *Understanding WAP*, Artech House, Boston, MA.

Voyager. URL: http://www.objectspace.com/products/voyager.

Z. Wang and J. Crowcroft (1996) Quality-of-service routing for supporting multimedia applications. *IEEE Journal on Selected Areas in Communications*, **14**(7), 1228–1233.

L. Wang and M. Hamdi (1998) HAMAC: An adaptive channel access protocol for multimedia wireless networks. *Proc. Seventh IEEE International Conference on Computer Communications and Networks*, pp. 404–411.

S. Weinstein, M. Suzuki, J. R. Redlich and S. Rao (1998) A CORBA based architecture for QoS-sensitive networking. *Proc. IEEE International Zurich Seminar on Broadband Communications*, ETH, Zurich, Switzerland, pp. 201–208.

Wireless Application Protocol Architecture Specification, WAP Forum, April 30, 1998. URL: http://www.wapforum.org.

Wireless Application Protocol Push Architectural Overview, WAP Forum, November 8 2000. URL: http://www.wapforum.org.

Wireless Application Protocol Wireless Application Environment Overview, WAP Forum, March 29, 2000. URL: http://www.wapforum.org.

Wireless Application Protocol Wireless Telephony Application Specification, WAP Forum, July 7 2000. URL: http://www.wapforum.org.

P. T. Wojciechowski and P. Sewell (2000) Nomadic pict: language and infrastructure design for mobile agents. *IEEE Concurrency*, **2**, 42–52.

W. S. V. Wong and C. B. Henry (1998) Performance evaluation of path optimization schemes for inter-switch handoff in wireless ATM networks. *Proc. ACM/IEEE MobiCom '98*, Dallas, TX, October 1998, pp. 242–251.

W. S. V. Wong and V. C. M. Leung (1999) A path optimization signaling protocol for inter-switch handoff in wireless ATM networks. *Computer Networks*, **31**, 975–984.

F. Wu Soft Handoff of Wireless ATM. URL: http://www.ctr.columbia.edu/~campbell/e6950/reprot_ifong.html.

C.-Q. Yang and A. V. S. Reddy (1995) A taxonomy for congestion control algorithms in packet switching networks. *IEEE Network*, **9**(4), 34–45.

R. Yuan and S. K. Biswas (1996) A signaling and control architecture for mobility support in wireless ATM networks. *Proceedings of IEEE ICC '96*, Dallas, TX, 1996, pp. 478–486.

Index